Sanitary Landfill

Sanitary Landfill

A State-of-the-Art Study

**National Center for
Resource Recovery, Inc.**

Lexington Books
D.C. Heath and Company
Lexington, Massachusetts
Toronto   London

Library of Congress Cataloging in Publication Data

National Center for Resource Recovery
  Sanitary landfill
  Bibliography: p.
  1. Sanitary landfills. I. Title.
TD795.7.N35 1974   628'.445   73-21693
ISBN 0-669-91744-3

Published simultaneously in Canada.

Printed in the United States of America.

International Standard Book Number: 0-669-91744-3

Library of Congress Catalog Card Number: 73-21693

Contents

List of Figures

List of Tables

Acknowledgments

The National Center wishes to give special thanks to Dr. Robert K. Ham, Department of Civil and Environmental Engineering, University of Wisconsin, for his significant participation in the preparation of this study. Appreciation is extended to Midwest Research Institute for an initial manuscript and to the individuals who reviewed this book and whose comments were invaluable in producing this final version: Professor Frank R. Bowerman, Director of Environmental Programs, University of Southern California, and Member of the California Solid Waste Management Board; Dr. A.A. Fungaroli, Enviro/Earth Ltd; and Mr. Ronald M. Vancil, Economic Consultant, The National Center for Resource Recovery, Inc.

Sanitary Landfill

1 Introduction

There is only one acceptable method for final solid waste disposal, and that is burial in the ground in a properly designed and controlled operation. Uncontrolled ocean dumping and open dumping on the land are obvious insults to the environment, result in nuisances and health hazards and in most cases are illegal. Incineration, composting, pyrolysis, separation for recycling and other intriguing but less known processes result in volume reduction with a residue which must be disposed; there will, therefore, be the continuing need for land disposal operations.

Sanitary Landfilling Defined

The traditional definition of sanitary landfill has been set forth by the American Society of Civil Engineers as: disposing of refuse on land without creating nuisances to public health or safety by utilizing the principles of engineering to confine refuse to the smallest practical area, to reduce it to the smallest practical volume, and to cover it with a layer of earth at the conclusion of each day's operation, or at more frequent intervals if necessary.[1] The American Public Works Association adds further that a properly executed sanitary landfill satisfies the following requirements:[2]

1. It minimizes vector breeding or sustenance by eliminating all possible harborage and food supply for rats, flies, and other vermin.
2. It adequately deals with the possibility of direct disease transmission, (e.g., trichinosis is controlled by not allowing swine to feed on infected garbage).
3. It effectively controls air pollution such as smoke and odor.
4. It minimizes fire hazard.
5. It minimizes the possibility of polluting surface and ground waters.
6. It effectively controls nuisance factors; i.e. the system must be aesthetically acceptable and noise must be held to a minimum.

In recent years the definition of the sanitary landfill has been broadened through common usage to include any controlled method of land disposal of solid waste that meets environmental, aesthetic, and public health standards. At the present time, there are two new methods which are gradually becoming recognized as meeting these high standards: milling (or shredding or pulverization) and baling. These methods will be considered in separate chapters in this report.

History of Development

Sanitary landfilling is a natural extension of open dumping, probably man's oldest form of planned refuse disposal. Similarly, open dumping is a natural extension of man's original form of disposal—casual discard of unwanted items onto the earth's surface. As man became more stable and built more permanent living quarters, it became increasingly difficult to wholly discard waste materials. The tendency to "hide" the refuse in depressions located some distance from his residence (and those of his neighbors) led quite naturally to setting aside certain areas for this purpose. Open burning may have been accidental in some places, but more commonly it came to be practical as a means of reducing refuse volume and prolonging the life of a site. The urbanization of society and increasing concern about environmental matters made obvious the unacceptability of the open dump, open burn operation, and thus new technology was developed.

Employing municipal refuse as a component of fill material in land reclamation projects was practiced several years before the term sanitary landfill was coined. Record of this practice dates as early as 1904 in Champaign, Illinois, but there it was devised as a means of reducing the amount of fill material to be transported to land reclamation sites rather than as a bona fide method of municipal refuse disposal.[3] Controlled tipping, as it is called in England, was developed as a method for solid waste disposal in England and Germany during the first World War. It is interesting that war conditions also promoted the sanitary landfill in the U.S., for it was during World War II that U.S. Army engineers engaged in extensive experiments in sanitary landfilling as a possible means of dealing with the rapidly rising solid waste volume at military installations. The publicized success of these experiments has been cited as the reason for the adoption of sanitary landfilling techniques by nearly 100 U.S. cities by the end of 1945. By 1960, more than 1,400 communities were reportedly practicing this method of refuse disposal.[4] However, subsequent investigation by the U.S. Department of Health, Education, and Welfare indicates that a good many land disposal sites in the U.S. do not meet true sanitary landfill requirements, and as recently as 1968 there may have been only 700 to 800 disposal sites in this country which could be classified as actual *sanitary* landfills.[5]

The situation has been improving since the 1968 survey, because the public acting through state regulatory agencies demands better solid waste disposal practices. Research and demonstration projects have provided technical foundations for proper design and operation of sanitary landfills and, through experience, many municipalities have learned that a well managed and widely accepted sanitary landfill is a reasonable and effective goal.

There are some situations where climate and the nature of available sites for landfill are such that it is difficult to economically operate a sanitary landfill disposal system, as defined by the ASCE. Starting in the late 1960s, variations have developed in the traditional sanitary landfill concept that provide useful alternatives in such situations. Milling and baling are the most widely publicized of these alternatives.

Purpose of this Report and Order of Presentation

The purpose of this report is to inform the reader about current sanitary landfill practices, both traditional and modern. It is not intended that the report cover each subject or system in detail; it is intended rather to provide a solid background on the entire subject, so that the reader can pursue additional readings, such as those cited here, on his own.

The remainder of the report considers planning and site selection (crucial to any landfill system), the operation and economics of the traditional sanitary landfill, milling and baling methods, public health, land use, and future possibilities.

Notes

1. Committee on Sanitary Landfill Practices of the Sanitary Engineering Division, *Sanitary Landfill*, ASCE—Manuals of Engineering Practices no. 39 (New York: American Society of Civil Engineers, 1959), p. 61.
2. American Public Works Association, *Municipal Refuse Disposal*, 2d ed. (Danville, Ill.: Interstate Printers and Publisher, Inc., 1966), p. 528.
3. *Ibid.*, p. 89.
4. *Ibid.*, p. 90.
5. *An Interim Report—1968 National Survey of Community Solid Waste Practices* (Cincinnati: U.S. Department of Health, Education and Welfare, October 1968), p. 8.

2

Planning and Site Selection

Hardly enough can be said for thorough planning before a sanitary landfill operation actually commences. A proposed landfill project can be assured of difficulties—or at best disappointments—when proper planning is neglected. Preoperational planning is necessary because a sanitary landfill project is an integral part of a community's total solid waste disposal system and of the natural environment where it is located. A mistake or unforeseen contingency in landfill execution could affect local tax dollars, public convenience, public health, and perhaps lives. Even the smallest landfill projects require adequate planning to ensure ultimate success.

Feasibility

The first phases of planning for a landfill project concern feasibility. It can be safely assumed that landfill disposal of any community's solid wastes is technically and physically possible today; so the question becomes that of the economic and political feasibility of transporting solid wastes to a suitable fill site and of meeting all of the expenses and restrictions associated with operating the site. The fact that solid wastes from places as dissimilar as New York City's Borough of Manhattan and the City of Mandan, North Dakota have been successfully landfilled attests to the wide range of municipalities which can consider sanitary landfill disposal. Factors that most affect the economic feasibility of a landfill project are: (1) the availability of a suitable site at reasonable cost,[a] (2) the distance that refuse must be hauled, (3) the availability of cover material (either on the site, or within economic haul distances at reasonable purchase price), (4) local wage rates, (5) cost of equipment, and (6) pre-fill and post-fill steps which must be taken to protect the surrounding environment and to enhance the final usefulness of the site.

Solid Waste Volume and Composition

While analyzing these factors, a community considering sanitary landfill disposal should evaluate the present and future solid waste volume to be handled

[a]Reasonable cost of land applies either to the cost of purchasing or leasing a disposal site.

by the project. A new housing project, a proposed shopping center, or a new factory may greatly affect the future volume of solid wastes and in turn affect the life of an existing or proposed fill site. Knowledge of refuse composition will allow greater awareness of the degree to which refuse can be compacted with available equipment and, consequently, a better projection of area requirements. Particular attention should be given to seasonal variations in refuse composition; summer months may yield a high percentage of tree trimmings that are difficult to compact. The composition of wastes handled at a landfill is usually controllable to some extent through advance planning as to what types of solid wastes will be acceptable—residential, commercial, industrial, demolition, abandoned autos, agricultural. The type of wastes accepted is important in establishing equipment and personnel requirements and projecting the life of the fill. The type of refuse accepted may determine who should be allowed to dump at the disposal site. From the planning standpoint, it does not matter who is allowed to dump— private parties, city compactor trucks, company owned trucks bearing industrial wastes, etc.—as long as the planned facilities and connecting arteries can handle the traffic patterns without congestion and without creating nuisances to the nearby residents.

Various rule-of-thumb techniques have been devised for projecting land area and volume of space requirements for landfill projects. According to the American Public Works Association, a landfill space requirement of two cubic yards per person per year is a reasonably conservative estimate. (This assumes a waste generation rate of 5.5 pounds per person per day and a refuse density of 1,000 pounds per cubic foot in the landfill. Two cubic yards of space per person per year is equivalent to 1.25 acre-feet per 1,000 people per year.)

More sophisticated methods of projecting area requirements should be used for detailed investigations. The most exact estimate of the waste to be handled at the site should obviously be used. Waste generation rates vary widely depending on many factors—such as climate, locale, season, etc.—and national average figures, such as the 5.5 pounds per person per day figure, should be used with caution. The density of compacted solid waste in the fill will vary depending on the degree of compaction provided and, very importantly, on the moisture content of the solid waste. A solid waste density of 1,000 pounds per cubic yard is common for reasonably well compacted shallow landfills (using solid waste of average composition and moisture content) up to 20 feet deep. For deeper landfills with three or more 20–foot lifts, the weight of the overlying refuse further compacts the solid waste to an average density of 1,250 pounds per cubic yard or more.[1]

The cover soil will, of course, take up some of the landfill volume. Normally one or more parts of soil for each four parts of compacted solid waste by volume is required daily to cover the waste materials completely. Thus, it may seem that at least 20 percent of the fill volume will be taken up by cover, and the capacity of the site for solid waste must be reduced accordingly. In practice, it is observed that the majority of the cover soil sifts downward into the voids in the solid waste, so that the actual loss in site volume as a result of daily cover may be only five percent.[1] The volume requirements for milled and baled refuse landfills are less than those cited above and will be considered separately in the chapters on these methods.

After establishing the size of a fill site required to handle a community's solid wastes for a given number of years, the search for an appropriate site or sites can begin. Or perhaps local officials may already have a site in mind and wish to estimate how long it could handle the community's waste load. In either case, an expected life projection for a given landfill location should be made early in the planning stages in order to facilitate future planning efforts and to make certain that adequate sites have been chosen to meet current needs. Some cities, notably Los Angeles, have planned for landfill locations as much as twenty years in advance. Most cities should plan at least ten years in advance in terms of population and waste volume when locating and purchasing fill sites.

Site Requirements

Haul Distance

In evaluating the feasibility of land disposal of solid wastes, it is essential that all costs associated with each of the various alternative methods and sites be compared, even though certain of these costs may not appear to be directly related to disposal *per se.* For example, it is commonly true that the additional cost to the collection system of transporting the solid waste to a more remote site is greater than the potential savings to the landfill operations through lower land costs. Similarly, when comparing sanitary landfilling with a central processing facility such as an incinerator, the cost of landfilling must include any extra hauling costs to the landfill, including any additional road repair and maintenance costs, etc., whereas the incineration costs must include residue and non-combustible landfill costs. It is common to learn through careful cost analysis that the cost of hauling solid waste to a landfill is greater than the cost of landfilling the waste.

A proposed landfill site ten to fifteen miles from the farthest point of collection can usually be traversed economically by standard size municipal

collection trucks. If the fill site is twenty or more miles from the city it is to
serve, consideration should be given to establishing one or more transfer stations
where refuse is loaded into large trailers of 75 cubic yards or more capacity.
Haul distances of fifty to 100 miles are usually beyond consideration for high-
way vehicles. In certain instances hauls which appeared too long to be feasible
have been made economically attractive by compressing the waste (e.g., dense
compaction, milling, or baling) or by using specially assembled transportation
systems, such as rail or barge. The future of sanitary landfilling as an economically viable disposal technique for large metropolitan areas will undoubtedly depend on new developments in the pre-fill processing and the transportation of highly compacted refuse to distant disposal sites.

Cover Material

The availability of cover material is another important planning consideration in selecting a sanitary landfill site. The best type of soil for landfilling is
a sandy or silty loam containing rocks no longer than six inches in diameter.
High clay content soils generally make poor cover material. Although they can
be used, one can anticipate operating difficulties in wet weather and surface
cracks in subsequent years. It is preferable to obtain a site where adequate cover
material is at the location. Planners can anticipate needing cover material
equalling at least one-fourth of the volume of compacted refuse. Transporting
cover material over long distances can be a costly proposition; however, sometimes it can be justified in fill areas close to population centers when the value of
the property will be substantially improved or the haul distance substantially
reduced as a result of the landfill. If sufficient cover material is not available,
land disposal operations using milled or baled refuse should be considered.

Other Considerations

There are many other criteria applied to landfill sites which vary from state
to state. In many states, zoning and the political location of the sites are of
crucial importance. Other factors may include the depth to bedrock and whether
the bedrock is fractured or can otherwise readily transmit leachates. The depth
to ground water, direction and flow and use of ground water, and the types of
soils which must be traversed by leachates in reaching and moving with ground
waters are important. Some states specify the minimum distance to the nearest

well or residence, or to surface water bodies, flood plains, heavily travelled highways, etc.

A thorough study will inevitably involve properly logged core sampling at several locations on and near the disposal site. Information obtained from these samples will allow for more accurate projections as to the amount of available cover material at the site, but it will also permit a better understanding of underground hydrography. By locating aquifers and rock strata, judgment can be passed on the location's suitability for landfill in terms of potential water pollution. At some sanitary landfill locations, explosive methane gas has moved through fractured rock or porous soil strata toward nearby homes. Potential tragedies can be substantially eliminated by core sampling and proper venting of such fills.

In summary, there are many factors that need to be considered in evaluating a landfill site. Community interests are protected by regulation in some states, or more generally by thoughtful design of any landfill operation.

Public Attitude

The most serious problem in acquiring landfill sites is invariably obtaining the acceptance and support of the community. All too often, sites ideal from engineering, economic, and regulatory points of view have been lost because of adverse public pressure. Drawing from the experience of landfill managers throughout the country, there seem to be several points which can be made in working with the public:

1. A crash public relations effort built around a specific site is frequently unsuccessful. A long-term public relations program designed to educate the public on a continuing basis is the most effective way of obtaining general public support. When specific prospective sites are announced, there will still be opposition, but hopefully the opposition will recognize not only the need for a site, but the true implications of the site in question. The opposition will be arguing the merits and potential nuisances of that particular site and will hopefully not be arguing the need for sites in general or using emotional arguments fed by rumors and false statements.

2. Crucial in building general long-term public support is the need for maintaining the best possible existing operations at all times. If the opposition can find negative features in the existing site(s), especially if these features could have been minimized by proper operations, they may have very

valid reasons for their opposition. Conversely, if the opposition has to go elsewhere to find flaws in landfill operations, they may be hard-pressed to enlist widespread support.

3. As a prospective site is announced, describe also a plan of operation and final use of the site. In most cases, proper design can result in a completed landfill which is beneficial to the community. The presentation of a well-conceived plan with sufficient flexibility to adapt to sound public requests, along with a definite time schedule and some means of ensuring the public that the plan will in fact be followed, can do much to gain public acceptance.

Engineering Planning

In order to ensure that the public has good reason for accepting a new sanitary landfill, adequate engineering planning is required. A sanitary landfill is, after all, an engineering project. As with any such undertaking, the assistance of competent professionals should be sought if such talent is not available within the organization planning the project. Surveys of proposed sites should be conducted and evaluation maps prepared. Recommendations should be offered as to construction and location of all-weather access roads, depth of fill at various locations on the site, location and amount of cover material at the site, required grades and culverts to permit proper drainage throughout site operations, equipment storage buildings, sanitation facilities, utilities, water supply for fire and dust control, and—most importantly—the type of sanitary landfill best suited for the particular location under examination.[b] A logical sequence of operations should be developed by the engineer along with site specifications at various stages of completion. Contingency operations for equipment or weather problems, etc., must be set forth in a good site plan.

Administration and Control

Once it has been determined that sanitary landfilling is politically, technically, economically, environmentally, and aesthetically possible for a particular location, it must be established who will administer and operate the fill. This may, of course, be established before the fill location is sought—for instance, when a private contractor approaches a city with the proposition of disposing

[b]Types of sanitary landfill are: area, trench, ramp, milling, and baling methods. These are discussed in Chapters 3, 4, and 5.

its solid wastes by sanitary landfilling. Nearly every type of administrative and operational arrangement has existed: city owned and operated, county owned and operated, city or county owned land and operation by private contractor, privately owned and privately operated, etc. In the history of sanitary land-filling, no one arrangement has stood out as being obviously preferable. What is important is that a community considering landfill disposal investigate the available alternatives and that it weigh them against each other in terms of total disposal costs, effectiveness, efficiency, environmental impact, and public health and safety.

The planning job is not complete until some type of control mechanism is outlined to ensure that the landfill is operated according to the highest stan-dards, that it is being filled according to schedule, and that it does not become too costly. Experience has indicated that one of the most difficult aspects of managing a good sanitary landfill is maintaining high quality operations day after day. Adverse weather conditions can make it extremely difficult to run a good sanitary landfill even if the design of the landfill provides contingency plans for such days. The person with ultimate responsibility for site operations must be qualified not only in a technical sense, but he must also have the power to make decisions quickly and he must have the time to work closely with his men and the site. There is a definite trend toward special training of landfill personnel. It is now generally accepted that the man who has run a tractor for years in road construction will not necessarily be a good man on the landfill. In-house training by experienced personnel and, more recently, special training courses offered through state, federal, and other agencies, are being used to provide such training. Well trained men and a good administrative system will do much to ensure a high quality sanitary landfill operation on a continuous basis.

An important aspect of controlling a landfill project is the keeping of accurate records of the amount of refuse deposited at the site. This data will, among other things, update projections of the life of the site and will aid in cost accounting for the disposal and, perhaps, the collection systems. If the size of the project can justify the purchase or leasing of scales, it is highly recom-mended. The weight of refuse unloaded has proven to be a much more effective measure for control purposes than volume. This issue becomes particularly critical when the landfill is operated by a different individual or agency than those who are dumping at the site.

Another important element of overall control of a sanitary landfill project is a meaningful and accurate accounting system. In an effort to bring about standardized accounting practices for landfill projects, the Bureau of Solid Waste Management in 1969 published a document entitled "An Accounting

System for Sanitary Landfill Operations." It has succinct, yet comprehensive, recommendations for an adequate accounting system; this document is reprinted in Appendix A.

In the event that a municipality or local government agency contracts for the services of a private organization to dispose of solid wastes through sanitary landfilling, most of the operating details will be controlled by the private contractor. To ensure that costs of disposal do not become exorbitant, and that the landfill will not be hazardous to health, it is a common practice for an operations agreement to be executed so that the community can maintain some control over the manner in which wastes are disposed. A model sanitary landfill operation agreement is exhibited in Appendix B.

End Land Use

An aspect of planning which takes on varying significance with different locations is the ultimate land use of the finished project. On occasion, end land use has been a prime reason for initiating a landfill disposal project. This has been the case in several land reclamation projects in marshy areas, in abandoned quarries and mines, and along tidal basins. In remote areas, land use is less of a planning consideration. However, at any sanitary landfill site, it is necessary that planners devote some attention to final vegetation cover in order to prevent erosion, and that arrangements be made for periodic inspection after completion of the project to ensure that settling or other problems have not made difficult the effective use of the completed landfill. Consideration must also be given to safe control of the movement of the gases from decomposition.

Notes

1.   American Public Works Association, *Municipal Refuse Disposal* (Danville, Ill.: Interstate Printers and Publisher Inc., 1970), pp. 94, 99–100.

3 Operation of a Sanitary Landfill

After sufficient planning and selection of a suitable site, the machinery for execution of a sanitary landfill can be set in motion. As indicated in Chapter 2, the type of landfill used at a particular location depends, for the most part, on the characteristics of that site.

Types of Sanitary Landfill

The three general categories of sanitary landfill considered in this chapter are: (1) the area method, (2) the trench method, and (3) the ramp method. The landfilling of solid waste processed by milling or baling is considered in subsequent chapters.

Area Method

This type of sanitary landfilling (Figure 3-1) is usually employed on sloping land, in ravines, canyons, marshes, quarries and other natural or man-made depressions. Refuse is dumped in or adjacent to the fill site and spread and compacted by a bulldozer or similar piece of equipment. Cover dirt is frequently obtained from a nearby high point and moved to the fill by a scraper, front-end loader, or bulldozer. Occasionally, cover material is hauled to the fill from a more distant location by dump trucks. If cover material is obtained from a point below the working face of the fill, drag lines have been used successfully, especially when the slope is too great to allow bulldozers to operate from the bottom of the fill upward (see ramp method), or when the area is wet and swampy. At least six inches of compacted cover material is recommended on the top and exposed sides of the compacted refuse. In practice, nine inches or so of cover is commonly required to cover the solid waste completely. This protective earth cover is laid at the close of each day's operation, or more frequently as refuse volume and operating conditions dictate. Once the compacted refuse is completely covered, it becomes a cell. The depth and area of such cells vary widely under different circumstances. The size of a cell is also

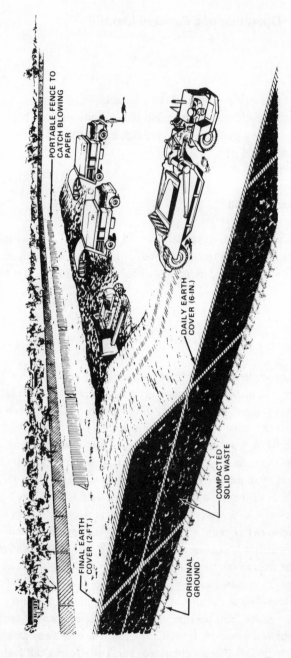

Figure 3–1. Area Method of Sanitary Landfilling. Source: T. Sorg and H.L. Hickman, *Sanitary Landfill Facts*, U.S. Department of Health, Education and Welfare, Bureau of Solid Waste Management, Pubn. no. SW–4ts (1970).

related to the number of vehicles unloading refuse at a given time. If many vehicles are to be expected at a given site, the width of the working face should be increased to accommodate more vehicles unloading side-by-side. This practice will minimize delays at the landfill, allowing the trucks to return to their collection duties more quickly. On the other hand, if the working face becomes too long, it becomes unwieldy and a potential nuisance. A general principle is to keep the working face as small as practically possible.

Generally, the deeper the cells, the greater the degree of refuse compaction which can be achieved. Refuse cells are usually constructed on top of previously laid cells. It is often advantageous to have refuse trucks pass over completed cells as they approach for unloading. This practice has yielded twenty to forty percent greater compaction, which reduces settling of the fill site in following years. Wet weather occasionally interferes with this practice, and thus an alternative dumping site adjacent to an all-weather roadway should be available. The final cover for an area landfill should consist of at least twenty-four inches of compacted earth. If trees are to be planted on the fill, the final cover should be at least thirty-six inches.

Trench Method

This method (Figure 3–2) is generally used on flat or gently sloping land. It involves the excavation and filling of successive parallel trenches separated by a three- to four-foot dirt wall. Usually dirt from the first trench is used to construct berms for windbreakers and remains as stockpile to cover the final trench. As dirt is needed to cover the first trench, the second is opened. Depending on topographic conditions and the volume of refuse, trenches vary in length from 50 feet to several hundred feet. Trenches should always be dug at least twice as wide as the tractors which must operate in them to permit easy maneuverability of the equipment. The depth of the trenches varies with soil and ground water conditions. A normal depth is eight to ten feet, although some trench-type landfills reach depths of fifty feet and more. The trenches are usually filled with cells six to eight feet deep, where the refuse is covered daily with at least six inches of soil to form cells. As with area landfills, at least two feet of compacted dirt is recommended as final cover. The equipment for trench landfills is often the same as that for area fills. However, the usefulness of a scraper-carrier may be offset by the value of a ripper in rocky soils or by a dragline in wet conditions or on steeply sloped banks.

EARTH COVER OBTAINED
BY EXCAVATION
IN TRENCH

DAILY EARTH COVER (6 IN.)

COMPACTED
SOLID WASTE

ORIGINAL
GROUND

Figure 3-2. Trench Method of Sanitary Landfilling. Source: *Sanitary Landfill Facts.*

Ramp Method

The third method of sanitary landfilling (Figure 3-3) is actually a variation of the area method. It is generally best suited to sloping land. Solid wastes are dumped on the side of the slope, spread, and compacted by a bulldozer. What makes the ramp method different is that cover material is generally excavated from below and in front of the working face of the fill as it is needed. The advantage of this method is that one piece of equipment (usually a bulldozer) can be used to perform all the necessary functions. Thus, the method is particularly well suited to smaller operations.

Compaction

Under normal conditions, refuse is compacted to about 1,000 pounds per cubic yard in all three methods of landfilling. This degree of compaction is achieved by spreading the refuse in a thin layer not more than two feet deep, and by making one to several passes over it with the heavy landfill equipment. Refuse is approximately three to four times as dense in landfills as it is in residential trash cans and nearly twice as dense as in compactor-collection vehicles. Several variables affect the density of refuse in a landfill, including the type of equipment used, method of operation, depth of the fill, and most importantly, the composition of the refuse. Of special concern in projecting the density of refuse in a landfill is the moisture content. Wide variations in fill density can be brought about simply by moisture content changes.

In order to achieve smooth operations and maximum refuse densities, it is frequently necessary to segregate certain items from the refuse as it is received. Tree segments, demolition debris, and bulky items, for examples, may be hauled to a different location (maybe at the same site) for disposal by burying or some other method. In some locations, tree limbs and trunks may be burned while a scrap dealer may be given old refrigerators, stoves, television sets, and auto bodies. Such items, which are difficult to compact, can be worked into sanitary landfills, but it may be necessary to provide special handling procedures so that they can be located at the bottom of deep fills or surrounded by special layers of dirt and thus be well compacted. Good compaction of all solid wastes at sanitary landfills is recommended to prolong site use and to prevent excessive settling and the creation of gas pockets and/or cracks in cover soils. Cracks allow surface water to flow directly into the refuse, and gas pockets may lead to fires and other problems.

Figure 3–3. Ramp Method of Sanitary Landfilling. Source: *Sanitary Landfill Facts.*

Special Operating Considerations

Certain operating procedures for a sanitary landfill have already been outlined in previous discussion. Refuse is dumped, spread, compacted, and covered.
Pre-fill excavation may be required, as with the trench method. The amount of
refuse deposited should be weighed when practicable. Traffic congestion can be
minimized and accessibility assured through proper road construction. Note,
however, that special attention must be given to the operation of a sanitary landfill when conditions are not ideal, as when adverse weather conditions occur.

Litter from Wind

In a 1959 survey of sanitary landfill operations by the American Society
of Civil Engineers, the operating problem most frequently reported was blowing
paper.[1] This problem is usually controlled by a combination of permanent and
portable fences. A permanent fence is recommended for the perimeter of the
landfill site. In addition to controlling litter from blowing paper, an outside
fence reduces the visibility of the operation. It also helps keep out unauthorized
scavengers. Paper accumulated against this outside fencing should be gathered
regularly and returned to the fill area for disposal. At many sites, ten- to twenty-
foot high fences mounted on skids are placed fifty to 100 feet downwind from
the dumping point and moved daily by existing equipment. The combination
of moveable and outer stationary fences will do much to reduce blowing
problems. As mentioned earlier, earthen berms can be constructed to serve
the triple purpose of blocking the view, controlling blowing paper, and providing a stockpile of cover material. Often fencing is used in conjunction with
dirt mounds to give added height to the peripheral barricade. Good design will
properly orient the working face in relation to the prevailing wind direction
to minimize blowing problems. Finally, roads should be constructed and operations designed so that trucks can dump in the most sheltered areas of the landfill
in high winds.

Winter Operations

Portable snow fencing may be required in northern locations to prevent
drifting snow from interfering with the operations. Severe winters may also
present the added problem of frozen earth. This is a particularly important operating problem for the trench method of landfilling, so it is recommended that

sufficient trenches be dug during the summer months to get through the winter without major excavations. Cover material should be stockpiled for winter use and covered with two to three feet of leaves or straw to prevent the dirt from freezing solid. Good equipment sheds will make it easier to start and service equipment, and cabs are available for heavy equipment to protect the operator.

Wet Weather Operations

Most earth moving operations, such as road building, cease during wet weather conditions. Unfortunately, the operator of a sanitary landfill does not have this flexibility, for refuse is not only produced daily, in all kinds of weather, but it must be collected, transported, and disposed daily as well.

Sandy soils present less of an operating problem in wet weather than do clay or silt soils. Arrangements should be made to stockpile cover materials which can be easily worked in wet weather. In periods of prolonged rain, it may be necessary to import cover material. Special wet weather dumping locations are suggested, and stockpiled sandy cover material should be convenient to these locations. These alternative locations must be situated so that landfill equipment does not get stuck when moving from equipment storage buildings to the working face, and they must be accessible by all-weather roads to refuse hauling vehicles. Many sanitary landfill operators have used solid demolition wastes (such as bricks, stones, and lumber) to lay temporary roadways to the working face of the fill during wet weather. This allows collection vehicles to move safely to and from the dumping point.

Every landfill should be constructed so as to maintain proper drainage in all weather conditions, both during operation and after completion of the fill. This may involve a series of dikes, culverts, ditches, and drainage pipes. Drainage is less of a problem on gently sloping land, or in high flat areas, but it often becomes a critical problem in low-lying areas and ravines. When a site is located on sloping land so that surface water from outside it flows naturally into or through it, dikes are usually placed at its upper side to divert water into drainage pipes or ditches around the fill area. In some instances, large permanent drainage pipes have been laid along valley floors beneath the fill to avoid interference with the natural flow of surface waters. Sometimes pumps are provided to remove waters ponded on the fill site itself.

In order to ensure proper drainage after the project is finished, a slope of at least one percent is normally required for the final surface of the fill. However, due to uneven settlement of most landfills, it is usually necessary to build in more than the minimum one percent slope in order to guarantee an end result of at least that amount. Extremely steep slopes for the finished face of a landfill should also be avoided to limit erosion problems.

Dust and Fire Control

At particularly dry times of the year, landfill operators should provide for proper dust control, especially near populated areas. This may be accomplished by drilling a well or connecting to a municipal water supply so that the working area can be dampened during the day's operation. Water may have to be imported by truck in more remote areas. Some landfill sites have constructed ponds or used existing ones near the work location as a source of water. An alternative to using water for dust control is spreading a light coat of oil or calcium chloride. Grass should be seeded on those areas at a fill site which are not scheduled for filling but which may create dust problems during dry months.

In addition to aiding in dust control, the availability of a water supply at a sanitary landfill site serves the cause of fire prevention and control. A properly maintained landfill operation is not likely to have fires; however, allowances should be made for the unexpected rubbish, brush or grass fires caused by carelessness or vandalism, or when embers are dumped. Fires can usually be extinguished by cutting off the oxygen supply with a layer of dirt, but water under adequate pressure should be available in the event fires spread to surrounding vegetation.

Fires have been known to develop in refuse cells that have already been covered. Underground fires are very difficult to put out. Water simply channels through portions of the refuse, having little effect on the fire in other parts. In such an event, the cell should be cut into with a bulldozer, spread out, and the fire extinguished with water or more probably with new soil cover. Normally, the dirt cell walls will prevent fire from spreading throughout the fill, but if a large portion of even one cell becomes burned, the reduction in refuse volume will result in uneven settling of the finished fill, cracking of cell walls, and possible spreading of the fire to adjacent cells.

Although not recommended, controlled burning is sometimes practiced at more remote landfill sites in order to reduce refuse volume. If such practice is allowed, it should be confined to large bulky refuse such as tree trimmings, and it certainly should be limited to only those items which will not give off an abundance of noxious fumes. Such burning is usually conducted at a pit or other appropriate site at the fill location, so that flying embers will not create fire risk. Besides the environmental hazards risked by burning at landfills, there is also the political risk of alienating the surrounding community—which in the long run may be a far greater risk than that to life and property. Note that any open burning practiced at a landfill site voids the proper use of the term sanitary landfill to describe the operation.

Salvage Operations

As with burning, salvage operations are not recommended at sanitary landfills. Even though recycling of solid wastes should play an increasing role in reducing the amount of wastes to be disposed, salvage activities must be considered secondary in importance at the actual sanitary landfill site, and thus be conducted at some point in the refuse disposal cycle away from the site itself. Any salvaging allowed at a sanitary landfill should be orderly and well supervised. Uncontrolled and unauthorized scavenging jeopardizes a disposal site's claim to being a "sanitary" landfill. If refuse is to be sorted at a landfill site, the sorting should be mechanized to the extent possible. Salvage activities should not interfere with either the traffic patterns of collection vehicles or the landfill equipment. In short, it should create no nuisances.

Post-operative Maintenance

Once the final earth cover is placed on a sanitary landfill and the area is made into a park or put to other use, responsibility for its maintenance does not end. Settlement and erosion from wind and water present the continual threat that the completed fill cannot be used to the maximum extent possible. Public ownership of the land provides some long term protection, but bonding may be desirable to insure proper maintenance where completed landfills remain in private ownership. Arrangements must be made for periodic inspection of the location for as long as twenty to twenty-five years after filling operations have ceased. Settlement is inevitably uneven, and this may present particular problems, depending on the site's end use. Leveling through grading or importation of additional cover material may be necessary.

Equipment Requirements

The selection of proper and adequate equipment is a key factor in the efficient operation of a sanitary landfill. The type, size, and amount of equipment required at a landfill location depend on the amount of refuse handled, the composition of the refuse, geological characteristics of the site, the fill method used, and other factors.

The most common piece of equipment found at a sanitary landfill operation is a track-type (crawler) tractor—often generically called a bulldozer. At small fill operations it is often the only piece of equipment needed because it can handle all the necessary functions—excavating, spreading, covering, and compacting. This machine can be made specially useful for conditions at a given site by equipping it with one of several different front-end attachments available: dozer blade, front-end loader, bullclam, bucket, scraper, or special trash blade. A representative track-type tractor and some of its various attachments are depicted in Figure 3–4. Wheeled tractors can be rigged with similar attachments to excavate and haul cover material or to spread and compact refuse. This machine is represented in Figure 3–5. The advantages of track-type tractors are durability and traction, more rapid and even compaction, predictable trade-in value. The advantages of wheeled tractors are speed, maneuverability, and the ability to move more readily from one location to another under their own power.

Past landfilling experience has suggested that the municipal wastes of cities up to 15,000 population can be handled by one tractor equipped with a bullclam (multi-purpose blade) or front-end loader of one cubic yard capacity.[2] Cities of 15,000 to 30,000 population need one unit of two cubic yard capacity, and cities of 30,000 to 75,000 population commonly have two tractors of three cubic yard capacity. However, when three or more units are required, it may be advantageous to equip them with different front-end attachments and have them perform specialized functions.

Based on the experience of landfill operation in Seattle, San Diego, and Los Angeles, one bulldozer tractor of 180 drawbar horsepower class and 47,000–pound gross weight can adequately handle about 250 tons of refuse a day, assuming that excavation of cover material presents no special problems and that cover is hauled no more than 100 yards.[3] However, small landfills for communities of 15,000 or less population, or those fills handling fifty tons of solid wastes per day or less can operate successfully with one tractor in the 10,000 to 30,000 pound weight range.[4] Heavier equipment (30,000 to 60,000 pound range) is recommended for sites handling the wastes of more than 15,000 people. Besides being able to handle larger volumes of refuse, heavier tractors provide much better compaction, resulting in more stable finished fills. Whenever a landfill project depends on only one piece of equipment, arrangements should be made for use of standby equipment during downtime or normal

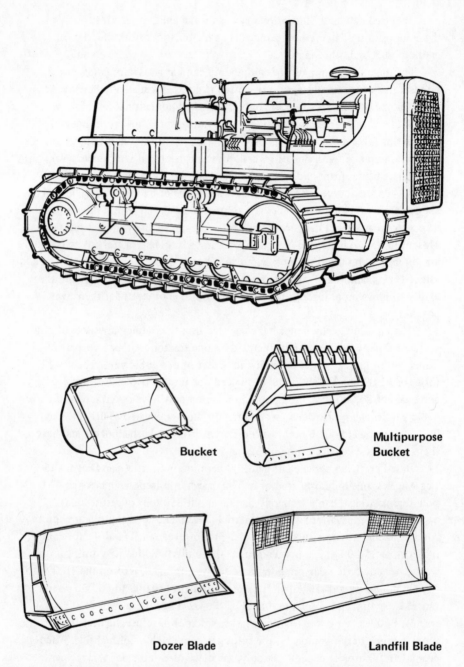

Bucket

Multipurpose Bucket

Dozer Blade

Landfill Blade

Figure 3–4. Crawler Tractor and Attachments. Source: *Sanitary Landfill Facts.*

Figure 3–5. Rubber-Tired Tractor. Source: *Sanitary Landfill Facts.*

servicing. Another alternative is to allow dumping at the site only one or two days during the week, thus freeing the tractor for other municipal services, such as snow removal and road repair.

When obtaining cover material presents special problems, landfill operators may wish to consider a different piece of equipment. A ripper may be necessary to loosen cover material in rocky or high clay content soils. When cover material is located over 100 yards from the working face of the fill and terrain permits, consideration should be given to use of a carryall scraper (Figure 3-6). For such a purchase to be justifiable, usually the landfill project must service communities of more than 30,000 people. Scrapers have been used to haul cover material as much as a mile to the working face of the fill. A prime consideration for use of a scraper for hauls of this distance is that of accessibility. Although rubber-tired and capable of traversing surfaced roadways, such heavy pieces of equipment would naturally create traffic problems if required to travel over roadways intended for public traffic, or even over on-site roadways intended for refuse truck use. So, for the most part, use of self-loading scrapers is confined to large landfills where the machine moves only within the confines of the landfill site.

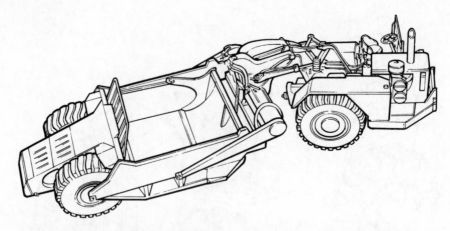

Figure 3–6. Self-Propelled Scraper. Source: *Sanitary Landfill Facts.*

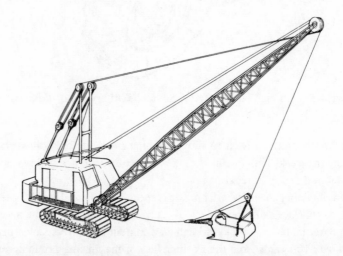

Figure 3–7. Dragline. Source: *Sanitary Landfill Facts.*

At large operations, a bulldozer is often assigned to work with a carryall scraper to assist in its loading operations.

Draglines (Figure 3–7) are more likely to be found at larger operations and, like the scraper, they are used for transporting cover material. The dragline is less maneuverable than a scraper, and its use is ordinarily limited to operations where large amounts of cover material must be transported up a steep incline or from underneath water. It may thus be practical to lease a dragline or to contract for the services of one during appropriate phases of the operation.

Figure 3-8. Steel-Wheeled Compactor Dozer. Source: *Sanitary Landfill Facts.*

The success and popularity of sanitary landfilling is evident from the fact that pieces of equipment have been specially designed for sanitary landfilling during the past few years. The most popular of these are wheeled compactors (Figure 3-8). Initially these units were little more than tire-type tractors with the rubber tires replaced by steel wheels, upon which were mounted breaker bars of some design. More recently these units have been designed with the spreading and compacting of refuse as the prime consideration.

Compactors apply extremely high pressures—through a combination of heavy weight and breaker bars or teeth mounted on the wheels—to break up, crush, and compact the solid wastes. Units weighing from 30,000 pounds to approximately 100,000 pounds are available. The steel wheels may be fitted with teeth that are rounded, pointed, or flat; in each case extremely high pressures are developed on the teeth and, in turn, the refuse. These units are especially useful on demolition and construction debris, and certain bulky items, including industrial wastes. They are quite maneuverable and have a weight distribution which is well suited to the slopes of a typical working face. They do not have sufficient traction to excavate cover dirt and are not particularly useful for applying cover because of the tendency of the teeth to poke through the cover and pull up or otherwise expose the refuse. Because compactors are not as versatile as crawler tractors, small sites using only one piece of equipment will normally choose a crawler tractor.

A list of modifications and auxiliary attachments needed to make heavy equipment more suited to sanitary landfilling has been compiled by the Solid

Waste Program of the Oregon State Board of Health.[5] These include such extras as heavy duty belly guards on tractors, reverse-flow fans to keep papers from adhering to engine radiators, meshed-wire spill plate extensions on top of blades, and extra-wide track pads. In addition, roll-bars are becoming popular in many landfills, and in California they are mandatory.

As with the design of the landfill, competent assistance should be sought when needed in equipment selection.

Personnel Requirements

The need for experienced and adequately trained personnel in the operation of a sanitary landfill has already been discussed. More is at stake than the safety of the crew operating the fill; the safety of the surrounding environment and the acceptability of the site to the community can be seriously jeopardized by a poorly operated landfill.

According to the American Public Works Association, the number of personnel required to run a landfill site varies from one site to the next. Factors most influencing the number of landfill workers needed are: the amount of refuse, type of fill, number of pieces of equipment, soil condition, and, to a lesser extent, climate and weather. Another factor which often affects manpower requirements is the number of organizations or people allowed to dump at a site. When only one contractor or only city-owned trucks dump at a city-operated site, fewer workers are required than at landfills where the public is allowed to unload refuse.

Most landfill operations handling the refuse of 15,000 or fewer people can be operated by one man.[7] For very small fills, such as in rural areas, part-time scheduled operation may prove economical and practical. An additional man is usually required to handle the refuse of communities between 16,000 and 25,000 population. Cities of 26,000 to 100,000 people generally use from 1-3 men at their landfill sites. Sanitary landfill operations serving heavily populated areas may require more operators, mechanics and groundskeepers. Every landfill operation should come under the general supervision (directly or indirectly) of qualified civil engineers. In addition to active involvement in the planning phases, engineering talent should be a part-time or full-time element of the landfill's actual operation.

Notes

1.   Committee on Sanitary Engineering Research, "Survey of Sanitary Landfill Practices, 13th Progress Report," *American Society of Civil Engineers,*

Journal of Sanitary Engineering Division, 87 SA4: 65–84, (July 1961).

2. *Municipal Refuse Disposal,* p. 113.
3. *Ibid.,* p. 114.
4. T. Sorg and H.L. Hickman, Sanitary Landfill Facts, U.S. Department of Health, Education, and Welfare, Bureau of Solid Waste Management, Pubn. no. SW–4ts (1970), p. 18.
5. W.B. Culham, "Equipment Needed for a Sanitary Landfill," *The American City,* 85(1) : 100, January 1969.
6. *Municipal Refuse Disposal,* p. 122.
7. *Ibid.,* p. 123.

4 Milled Refuse Landfilling

History

The practice of landfilling milled refuse, or shredded or pulverized refuse as it is sometimes called, began in Europe approximately twenty years ago. The earliest installations were used primarily to reduce landfill space requirements and to produce a refuse mass suitable for composting; however, observations of milled solid waste soon indicated that some disagreeable characteristics of the unprocessed waste were, for some reason, not so evident. Reports describing landfills containing milled refuse said, among other things, that the odor was diminished, that rodents and insects were not attracted to the material, and that the sight of milled refuse was not as objectionable to the public as unprocessed wastes. Further, these promising results were obtained without soil cover to mask or hide the refuse. The result was that milling gained in popularity over the years to the point that there are now approximately 150 installations in Europe where refuse is milled primarily for landfill disposal. This figure does not include installations which mill refuse primarily for composting or incineration. Note that these installations still landfill some milled refuse, however, as process residue or rejects, during downtime, etc.

In the United States the practice of milling began in 1966 when the City of Madison, Wisconsin, entered into a demonstration grant with the federal government and the Heil Company of Milwaukee. The Heil Company had observed milled refuse landfills in Europe, obtained rights to sell a mill of French manufacture, and wished to have the mill evaluated in the U.S. The University of Wisconsin was contracted to do the evaluation of the milling facility and the milled refuse landfill. This facility has been in operation since 1967 and now processes approximately 250 tons of residential and light commercial solid wastes per day for landfilling. Over the last few years, there has been increasing interest in milling, largely the result of conclusions drawn at Madison and in Europe, until there are presently over twenty such installations operating or under contract.

To Mill or Not to Mill

Milling solid waste for landfill purposes should be considered for each site or operation along with the various other methods of solid waste reduction and disposal available. In evaluating milled refuse for landfill disposal, it is instructive to look first at the five most common reasons cited by owners of existing facilities for choosing that method: (1) better acceptance by the public, (2) longer site life, (3) less need for cover, (4) better quality of daily operations, and (5) greater adaptability for additional processing, especially separation for resource recovery. These reasons will be discussed in more detail in the order presented.

Better Acceptance

Many operators of milled refuse landfills and regulatory officials have described how milled refuse has enjoyed better public acceptance than unprocessed refuse in a landfill. Of course, public acceptance is of special concern when a new site is being procured, and experience indicates that this is when milling will most likely be incorporated into a disposal plan.

The reasons for improved public acceptance of milled wastes as landfill are not well understood. In many cases, it may be easier to live with the thought that processed rather than unprocessed "garbage" will be hauled or disposed near one's residence. But there are more basic reasons which may be playing a role in reducing objections. Milled refuse does not look like the material the citizen throws away daily and finds objectionable. It looks like shredded paper for the most part, and as one moves more than perhaps 100 feet away from it, it looks like a gray mass with no distinguishable features. It is easily contoured to smooth, unbroken surfaces, and the contour is not readily disrupted by erosion. Less heavy equipment, traffic, and manpower is observed at most milled refuse sites, because most of the activity takes place in the milling plant itself and not on the landfill where it is open to view.

Longer Site Life

Examples of 10 to 75 percent extension of site life can be found in the literature on milling. Part of the reason for this wide range is the varied composition of refuse, ranging from relatively loosely packed paper or bulky wastes to dense wet materials. Another reason for the range in figures is improper testing methods, or even the lack of scientific testing methods. It is unfortunate that mere estimates of volume reduction have been accepted by some people as factual. A final reason for the discrepancies in volume savings is the various degrees of compaction applied at different landfill sites.

Data from Madison, Wisconsin, indicate a density increase by milling in equally well compacted milled and unprocessed refuse cells of similar composition, of 15 percent on a dry weight basis. Additional data indicate that as more compaction is provided, there will be less percentage difference in density between milled and unprocessed refuse. These data also indicate that the 15 percent figure is near the minimum savings which can be expected. With normal compactive effort (not as much as at the Madison test cells), field experience from other installations indicates a volume reduction of closer to 25 to 30 percent. Note that all of these figures refer only to refuse density, and any savings in landfill volume by the need for less cover on milled refuse must be added. Thus, although the exact figures vary, it is generally accepted that site life will be prolonged significantly with milled refuse, a factor which is of special importance to many landfill operators.

Less Need for Cover

U.S. experience thus far has corroborated experience elsewhere that a milled refuse landfill does not need daily soil cover, except possibly for aesthetic reasons, depending on the visibility of the site. Tests and observations to determine any need for daily cover have considered blowing paper and debris, dust, odor, public objection to the apppearance of the site, rat problems, fly problems, fire danger, gas production and dispersion, leachate production, and general operation of the site. There appears to be no common adverse observation requiring daily cover on milled refuse, but at certain sites specific circumstances may dictate daily or less frequent cover. If the site is very visible and close to residences or the general public, it may be useful to cover the site for aesthetic reasons (although at no site reported in the literature is this routinely done). Also, birds are frequently found on sanitary landfills, so such operations should include very frequent and thick cover, whether the refuse is milled or not. This is especially important when landfills are located near airports.

If anything but properly milled refuse, such as bulky or industrial wastes, is to be landfilled, this material should be compacted and covered daily with dirt or several feet of milled refuse to minimize fire potential and to improve the appearance of the site. Finally, if a site is so located that it would be desirable to prolong active decomposition and to minimize the volume of leachate production per unit time, the site should be covered at intervals, but not necessarily daily.

It is obvious that savings in cover for a milled refuse operation can be especially important if good cover has to be hauled in or excavated at premium cost.

Better Quality of Operations

One problem frequently cited by operators of sanitary landfills is maintaining high-quality landfill operations under adverse weather conditions. A particularly desirable characteristic of milled refuse is that road vehicles and landfill equipment can be driven over it easily when it is compacted and reasonably smoothed. This feature is of special importance in wet weather, for it remains easy to drive on milled refuse when most soils have turned to mud. The lack of exposed earth at a milled refuse landfill not covered daily with soil makes the site accessible in either wet or dry weather conditions.

Another weather problem faced by some landfills is extremely cold weather. Obtaining and handling cover dirt can be especially difficult at such times, so the decreased need for cover with milled refuse makes it easier to operate such a landfill in cold weather.

The final weather problem is litter blowing from a sanitary landfill on windy days. Because of the smaller size of milled refuse particles, and because they intertwine to form a network or lattice, properly milled refuse does not blow as readily as unprocessed refuse. The blowing which does occur is typically limited to short distances for paper particles, and to a rolling action which moves film plastics along the surface of a landfill until they are stopped by a low fence. If refuse is not milled properly, larger pieces of paper will blow. Experience suggests that 90 percent by weight of the particles should pass a three-inch sieve to achieve the best landfill characteristics.

Greater Adaptability for Further Processing

The final reason commonly given for milling refuse for land disposal is that this process can reduce the refuse to a mass of relatively uniform particle size so that it can be more readily treated by additional processing. The most obvious scheme is to provide magnetic separation of ferrous metals for recycling by the steel industry, which is done at many milling installations. Some installations also have air classifiers for further separation of certain solid wastes fractions for recycle. Paper and several pressed fiber products (e.g., particle board and packaging materials) have been made successfully from air classified solid waste in test programs, and it is likely that such processes will be operating on a production basis in the next few years. Other devices have been used to separate other solid waste fractions such as glass, non-ferrous metal, etc. A common feature of nearly all separation processes is that they require that refuse be milled prior to separation.

Other "advanced" treatment processes for solid waste also used milled refuse, including advanced incineration (fluidized bed or as a fuel supplement in coal-fired power plants), composting, and pyrolysis.

The conclusion is that as separation for recycling or other advanced processing becomes more popular, there will be more landfilling of milled refuse. Residues or unusable fractions of the milled solid waste will be landfilled, as will material milled but not processed due to downtime, etc.

One final point is that a milling facility may be centrally located and be in effect a transfer station as well as a facility processing refuse for disposal. A transfer station to compress refuse for loading into large vehicles for long-distance hauls to remote landfill sites is especially useful in large metropolitan areas where nearby landfill sites are unavailable.

There are three reasons which might be given for not milling refuse for land disposal: cost, administrative problems, and the newness of the method. Cost is, of course, the major factor of concern in evaluating a milling installation. A milled refuse land disposal operation will cost more than an unprocessed refuse-daily cover sanitary landfill in most locations. If site volume is not critical, and the site has sufficient cover material or cover is readily available, milling will be an expensive alternative to a traditional sanitary landfill in most situations. It is necessary to project carefully all of the costs associated with each type of operation and to attempt to take into account non-monetary factors such as public acceptance and quality of operation on a daily basis when determining whether to establish a milling installation. Additional comments on cost will be found in Chapter 6 on economics.

Another factor to be considered in evaluating milling is whether a processing facility can be managed in optimal fashion by the agency involved. Overtime policy, scheduling, incentives, maintenance programs, and the setting of production figures are among the considerations which need careful attention in operating a production facility. If a milling facility cannot be run reasonably well within a governmental agency because of the agency's policies, a decision to mill may be more difficult to justify.

The final negative factor to be considered is that milling is not yet an established method of solid waste processing and disposal, and thus such installations are predictably subject to publicity and scrutiny. Any lapses in operation, or negative features of the site will be publicized, which may arouse special concern among regulatory agencies and the public.

Equipment

A basic milling facility consists of a dumping area, feeding system, mill, and milled refuse transport system. The dumping area must be big enough to handle the maximum number of vehicles expected at any one time, and must have sufficient storage space for a two or three shift operation, downtime, etc. If the mill is fed by a conveyor system, the dumping area is typically a floor extending on one side to the feed conveyor hopper. Trucks can dump directly onto the conveyor or, more commonly, onto the floor, from which the solid waste is pushed by a front-end loader into the feed conveyor system. If the feed system involves a crane, the solid waste is pushed by a front-end loader into a dumping pit. It is difficult to feed evenly material as diverse as solid wastes into any processing facility, and equipment problems reported by milling facilities often involve the feeding system.

There are many manufacturers of milling equipment, and each produces a product of supposedly unique design. The most common type of mill is the hammermill, where rapidly rotating hammers strike the solid wastes until the waste material is sufficiently reduced in size to pass a given grate size opening. In all cases, it appears that the ground refuse should be sufficiently reduced in size so that approximately 90 percent of the refuse by weight passes a three-inch screen. Such a grind has generally resulted in acceptable landfills.

The milled refuse is usually transferred by conveyor to a wagon or truck which hauls the material to the landfill. The size of the vehicles is based on the production rate of the facility and the distance to the landfill. Note that the use of a large truck, barge, or other sizeable unit can bring about savings in hauling costs. At the landfill one piece of equipment for spreading and compacting is generally sufficient. A track-type tractor or a wheeled compactor is generally satisfactory for this purpose. Milled refuse is easily spread and compacted, which results in savings in actual landfilling costs when compared to costs at conventional landfills.

Landfill Design and Operation

The same care in designing and operating a landfill is required whether the refuse is milled or not. It is, perhaps, easier to design a milled refuse landfill in the sense that little or no cover is required and operational problems may be expected to be reduced; however, this by no means reduces the need for a well designed and operated processing facility.

There are two polar concepts which have been used in operating milled refuse landfills, plus many more concepts between the two extremes. One concept holds that the milled refuse should be spread in thin layers and well compacted to achieve maximum density. A density in a 10–foot deep landfill of 1.425 pounds per cubic yard (at 52 percent moisture on a dry weight basis) may be expected, according to results at Madison. Higher densities will be achieved in deeper landfills with many layers. The other extreme is to spread the milled refuse loosely in one- to two-foot layers and then let the material degrade for perhaps two months before compacting and covering the compacted layer with another layer of freshly milled refuse. Since no traffic is to be allowed on the uncompacted refuse, large land areas are necessary for this concept to be put into practice. However, it is likely that the finished landfill will be more stable and will undergo relatively little additional settling and decomposition. Note that any unprocessed refuse landfilled at a milled refuse site must be compacted and covered like any sanitary landfill. Milled refuse in layers several feet thick can be used for cover in this case.

Reasonable final uses for a milled refuse landfill are the same as those for conventional sanitary landfills. The lessened need for cover, the vehicle supporting capability, and the fewer problems due to weather and other operational concerns may make it easier to design a landfill with a unique end use in mind. If no unprocessed refuse is placed on the site, it may be expected that less differential settlement will result. The density of milled refuse is higher at the time of placement, which would also probably reduce settling and may make the site more desirable for some end uses. Although milled refuse supports much volunteer vegetation after a year or two, and after five years begins to look much like soil, it is best to cover the finished site with two or more feet of soil to finish the operation properly.

5 Baled Refuse Landfilling

History

The compression of solid wastes into bales by the application of high pressures prior to landfill disposal is a more recent development than sanitary landfilling with either unprocessed or milled refuse. The baling process has been used in the secondary materials industry for many years, where ease of handling and transportation rates have been motivating factors. It was not until about 1960, however, that the process was first used on a continuous basis, in Japan, to prepare solid waste for landfilling. It is therefore one of the most recent developments in landfill practice.

Baling of refuse has been considered on a test basis by several organizations. The most comprehensive test information and projections in the United States were developed at Chicago under an EPA grant in the late 1960s. Virtually every major city in the U.S. has considered long distance hauling of solid waste to remote landfill sites, and these plans have usually specified baling in order to utilize fully the weight-carrying capacity of railroad cars. Most of these plans failed to materialize because of economics or, perhaps of more significance, because of the inability to obtain sites, even at remote locations, as a result of public pressure.

There are three full-scale facilities known to be producing bales for landfill at the present time. They are at Cambridge, Massachusetts; San Diego, California; and St. Paul, Minnesota. The Cambridge facility uses a centrally located baler with a rated capacity of 1,600 tons per day to compress solid waste for long distance hauling by rail or flat bed truck. It is privately owned and operated. At San Diego, a continuous baler was obtained and is being operated and evaluated by the City under a federal demonstration grant from EPA. The San Diego baler is somewhat different from the other two, for a continuous bale is formed and pushed through an extrusion chamber with separator blocks inserted at 70- to 80-inch intervals. Wires pushed through slots in the blocks are used to bind the bales. The other balers form one discrete bale at a time. The St. Paul plant is owned and operated by the firm which manufactured the baler. In return for a dumping fee, the plant processes the refuse into bales, which are transferred by flat bed truck to a landfill. Additional baling plants are reported to be under serious consideration at several U.S. cities. The most notable baling efforts outside the U.S. are found in Japan, where several production facilities are reported to be operating.

To Bale or Not to Bale

Because baling is a recent development, it is more difficult to document the experience and the rationale for baling than for milling. However, the motives commonly given for baling by cities that have considered it are: (1) to compress the refuse for more economical transportation, (2) to make the solid waste less objectionable to the public both during transport and at the landfill, (3) to improve landfill operations, and (4) to increase landfill life.

Compress the Refuse for Transport

Solid waste disposal has always been held as one of the lower land uses by the public and, unfortunately, by many planners and public officials. Thus, as urban sprawl encompasses more and more land and puts it to residential, commercial, and a few public uses, it has become more difficult to find landfill sites in urban areas. The need for sites has caused most of our larger cities to consider long distance hauling of solid waste to remote areas.

Economical long distance hauling requires that the maximum weight of refuse be carried per vehicle, and one way to increase the density of refuse is to compress it. However, depending on whether the load limitation per vehicle is a volume or weight restriction, increasing levels of compaction may or may not be useful. In San Francisco, for example, refuse crushed by a crawler tractor is loaded into open top trucks, tamped if necessary with a power shovel, and the allowable pay load on a weight basis is invariably reached. In Milwaukee, stationary compactors are used to press refuse into enclosed trailers, and again legal road limits are normally by weight rather than volume, and further compaction is not necessary. Baling is thus one of several methods to reach the allowable weight limits on trucks.

With rail haul the situation is somewhat different. Now the volume rather than the weight of a refuse load is normally the restricting factor, so increased levels of compression, as with baling, are especially desirable. It is anticipated that long distance hauling will become more common in the future, and that baling will be a significant factor in such systems, especially when rail haul is involved.

Make the Solid Waste Less Objectionable

When compressed tightly into a bale, solid waste seems to lose much of its identity. Food wastes and other objectionable components of solid waste are hardly observable; and the bales are relatively odor, dust, and nuisance free. If additional precautions are desired, it is not difficult to enclose the bales in plastic or some similar material, to cover them with canvas during shipment, and/or to cover the bales with soil in the landfill. The experience with baling operations thus far has been good as far as public acceptance is concerned. Acceptance may be even better if the refuse is shredded prior to baling.

Improve Landfill Practice

Although landfilling with baled solid waste is a relatively recent practice, experience with bales thus far indicates that many of the operational problems in a landfill can be reduced substantially. Blowing paper and other debris is virtually nonexistent, and trucks can drive over the bales in wet weather.

There is less activity at the site, for the refuse need not be spread and compacted and is necessarily confined to a relatively small working face. No odor, fire, and vector problems have been reported, and the acceptability of a baled landfill is undoubtedly not as dependent on daily cover as a conventional sanitary landfill. Tests are underway to document the characteristics of a baled refuse landfill, and additional tests are planned by EPA. Preliminary indications are positive.

Increase Site Life

Bale densities of 1,600 to 1,800 pounds per cubic yard (wet weight) are commonly achieved. At San Diego the average well-formed and tied bale is approximately 1,690 lbs/yd^3. This density must be reduced somewhat because of voids between the bales as placed in the fill and because of the use of cover, so the *effective* density in the sanitary landfill (pounds of refuse per cubic yard landfill volume) is lower than the above density. At San Diego a survey ending in May of 1973 determined that the overall density is 1,510 lbs/yd^3 wet weight. Although this figure is subject to change depending on additional landfill testing, it is apparent that site life will be increased substantially with bales. Using 1,000 pounds per cubic yard for a shallow unprocessed refuse sanitary landfill, ranging up to 1,250 pounds per cubic yard for a deeper landfill, the volume savings would range from approximately 20 to 50 percent. The compressive effect of many tons of material placed on unprocessed refuse in a deep landfill will be larger than with baled refuse.

The three arguments against milling given in Chapter 4 should be repeated here. They concerned the costs, the administration and control, and newness of the method. Baling does represent a cost in processing solid waste prior to landfill disposal, and this cost must be justified on the basis of reduction in hauling or disposal costs, better site operations, public acceptance, etc. Further because it involves a processing facility, management of a baling operation should be set up in such a manner as to utilize the best principles of accountability and production to be found in industry. Special treatment and concern by the public and professionals alike will be focused on a new concept, such as baling. This concern is a good thing, but the persons involved with a baling facility must be prepared to be especially careful to run an acceptable system at all times.

A final potentially negative feature of baling is handling the bales quickly and efficiently. Handling problems appear to have been minimized at San Diego with a rubber-tired front-end loader equipped with a basket, and at St. Paul by use of a fork lift, but there is concern that problems may be encountered in this area, especially under adverse weather conditions.

Equipment

A baling facility consists of a dumping floor, feeding system, baler, and bale conveyance and transport system. The dumping floor is much the same as in a milling facility, where trucks may dump either into a pit or onto a flat surface. If a pit is used, a crane will move the solid waste from the pit to the baler. If the baler is fed by conveyor, trucks may dump into the conveyor hopper directly or, if the hopper is full, onto the floor, from which an end-loader can push the refuse into the hopper as needed.

There are several types of balers available. The single-stroke, continuous operation baler forms one long continuous bale which is cut into small segments as appropriate for handling. Other balers form one bale at a time and may involve one, two, or three strokes. In the three-stroke machine, the refuse is compressed from all three perpendicular directions in sequence. In a two-stroke unit the refuse is compressed from two perpendicular directions, etc. Some persons feel that their baler operates best with milled or pulverized refuse, others say that such processing is an unnecessary expense. Most agree that a bale of milled refuse looks better than one of unprocessed refuse, but there is a question whether the improvement is worth the added expense. Other decisions faced by the designer of a baling system are whether to tie the bales (reported to be necessary if the refuse is milled prior to baling), and whether the bales should be wrapped. These decisions will depend on the particular baler and the method and characteristics of the bale handling, hauling, and disposal systems.

Bales will generally weigh several tons apiece (1½ to 2½ tons at St. Paul and San Diego), and thus special equipment to convey them to the landfill must be procured. Roller conveyors seem to have worked well for transporting the bales from the baler onto flat-bed trucks, which are themselves fitted with roller conveyors. Overhead cranes have also been used to haul bales to flat-bed trucks or rail cars. The St. Paul plant has an automatic loader which loads sixteen 3,000-pound bales directly onto a flat-bed truck. The vehicles may be unloaded by crane, fork lift truck, or direct dumping and, as mentioned previously, fork lift front-end loaders have been used successfully for moving the bales at the landfill into place at the working face of the fill.

Landfill Design and Operation

It is difficult to draw guidelines for a baled refuse landfill because few of them have been operated. Experience thus far indicates that baled refuse landfills offer relatively nuisance free operation and are not prone to rodent or insect infestations, fires, odors, blowing debris, etc. Further, if the bales are arranged to form a flat surface which can be used by vehicles, traction appears to be good even in wet weather. No information is available on decomposition of bales, with its attendant leachate and gas production, but it may be expected that a properly constructed bale landfill (minimum voids) will not be likely to settle much, and the differential settlement should be negligible. There has been some publicity about placing bales in water to fill depressions or stabilize shore erosion, etc. It is not yet known whether such uses will in fact prove to be constructive, or whether water quality will be impaired.

6 Economic Considerations

There is perhaps a wider range of variables involved in the cost of a sanitary landfill than in other forms of solid waste disposal. This is true because every sanitary landfill project is unique; it is impossible to specify detailed requirements as to equipment and operating procedures which will apply to each and every landfill project. The cost of land may represent a substantial part of the initial investment, but in terms of total disposal cost, land may represent only a few cents per ton of refuse disposed. Operating costs is where landfill projects are most likely to run into financial difficulty.

The total costs of sanitary landfilling with unprocessed refuse have been reported by various sources as ranging from 50 cents per ton of refuse to over $4 per ton. Generally, these costs compare favorably to other average costs associated with municipal refuse handling and disposal—collection and hauling ranging from $12 to $20 per ton and incineration from $8 to $15 per ton. Most of the wide range in reported landfill costs may be explained by varying conditions (such as wage rates, equipment requirements, and land cost) at different locations. However, in comparing landfill costs, the fact should not be overlooked that different reporting systems and accounting practices may tend to obscure the true total costs for some operations.

Initial Investment

Land

As previously indicated, the availability of a suitable fill site is frequently the most pressing issue faced by sanitary landfill advocates, especially around large metropolitan centers where property values are high and refuse volume great. New York City, long a practitioner of sanitary landfilling techniques, has recently been trying to phase out landfilling of refuse simply because of the scarcity of land. Some existing fill operations around metropolitan New York are being reserved for the disposal of incinerator residue.

In more remote locations, land may often be obtained for a few dollars an acre, and land costs may average as low as 1 cent per ton of refuse disposed.

At Los Angeles, land prices have exceeded $27,000 per acre for landfill opera-
tions, but even at these prices, land costs are only 10 cents per ton of refuse for
100-foot fills and still lower for deeper fills. Thus, with total landfill costs ranging from 50 cents to $4 per ton of refuse, it is apparent that land costs are a
relatively small portion of total disposal costs—usually ranging from 2 to 10 percent of total costs.

It follows from the rather small percentage of sanitary landfill costs
associated with land expense that the hauling expense involved in using different prospective sites may be of overriding importance. Land costs must
normally be substantially lower for a more remote site to offset increases in
transportation costs to the site.

In an accurate accounting sense, the value of the land after completion of
the fill must be counted against the original cost of the land. In many instances,
the value of the property increases after filling, in which case the land ultimately
costs nothing (if it is sold) or its value as a park or recreational facility more
than offsets the purchase price of the land. The possibility of leasing land for
sanitary landfill disposal should not be overlooked. Lease agreements are frequently quite favorable for the lessor, especially when the condition of the land
is improved—as with abandoned quarries and strip mines. Four communities
in northeastern Pennsylvania jointly leased sixty acres of abandoned coal strip
mines for $1 per year, enabling the land, which was nothing more than a worth-
less eyesore, to be turned into rolling grass-covered hills through sanitary landfilling.

Planning

These expenditures must be regarded as an investment and charged to total
disposal costs. As with land costs, some portion of planning costs may be
absorbed by parks and recreation agencies, but the planning dollars spent on
geological sampling, operational procedures, environmental protection, special
surveys of solid waste volume, composition, and landscaping should be allocated
to refuse disposal. As with land costs, planning expenditures are ultimately a
small portion of total disposal costs, and planning is certainly no place to cut
budgets. Every dollar wisely spent in planning will be more than justified in
terms of quality of the finished project and smooth operations during filling.

Site Preparation

These costs are usually related to land costs and planning (site selection)
costs. By conducting a thorough search for the best available site, and by spend-

ing the money to obtain that site, preparation costs can often be reduced. Hasty selection or purchase of the cheapest available land may be false economy if pre-fill preparation expenditures are extensive. Some preparation expenditures—such as for fences, signs, access roads, and utilities—are common to all fill sites. However, clearing trees and brush, excavating cover material, and installing drainage ditches, culverts, and leachate and gas control devices are areas where some landfill projects run into unexpected outlays. As with planning, site preparation is not a phase of the project to be stinted, because repairs on poorly prepared sites can be more costly in the long run than adequate site preparation. Some large fill sites have required expenditures of over a million dollars on planning and site preparation alone; however, this is the exception rather than the rule. For most fill projects, site preparation represents a very small portion of total disposal costs, but it does represent an investment and should not be neglected in planning and budgeting.

Facilities

These expenditures are most often directly related to the size of the operation, and to a lesser extent the climate. A small fill may have no structures, but some type of building is usually recommended to provide sanitation facilities and shelter from inclement weather. Equipment storage and maintenance are usually necessary if machinery is to remain at the site. Some type of storage tank for fuel is required; this should be conveniently located, but not so as to create hazards in the event of rubbish or grass fires. A large fill operation might have separate structures for administration, equipment, sanitation, and weight scales. However, larger operations can usually afford more elaborate facilities, because of larger volumes and resulting low facilities costs per ton of refuse handled.

Equipment

Equipment represents a substantial initial investment for the authority operating a sanitary landfill if it is purchased rather than leased. An analysis can be conducted to determine the most economically favorable arrangements—leasing or purchase. Resale or scrap value of equipment should be taken into consideration when allocating this investment expenditure to the amount of refuse received.

The amortized investment and operating costs of equipment are frequently stated together as a single figure per ton of refuse or per hour of operation. This is done to arrive at a total cost figure per piece of equipment, so that the overall efficiency of different types of equipment can be compared. This method of arriving at equipment costs will be discussed further in the "Operating Costs" section of this chapter. It should be pointed out, however, that landfilling equipment is expensive and, when purchased, represents a substantial portion of the total initial investment.

Equipment dealers should be contacted to determine exact prices of the machinery needed for a particular fill operation. However, a rough rule of thumb for estimating equipment prices is a dollar per pound. Thus, a 20-ton tractor will cost in the neighborhood of $40,000. And a landfill requiring two tractors and a scraper will probably necessitate an equipment investment of $120,000 to $150,000.

Refuse Processing

Refuse processing by either milling or baling can add substantially to the initial investment in a sanitary landfill. If the processing equipment is related to a specific site, the equipment must be amortized over the life of the site or the life of the equipment, whichever is deemed less, and possibly even a dismantling cost should be projected and prepared for. If, however, the processing facility is to be operated in conjunction with several landfill sites, and is in effect a transfer station, it may be useful to consider the cost of processing and hauling the processed material as the third major cost generator—to be added to collection and disposal (landfill) costs to give the total cost of solid waste management.

A rule-of-thumb figure for the capital cost of an enclosed refuse milling facility is $4,000 to $6,000 per ton of rated capacity per shift. This cost would include the total costs of the building and foundation, dumping floor, refuse feeding system, mill, refuse transport system to a nearby landfill, and operating facilities. There is an economy of scale with multiple mill installations, in that the added cost of each additional mill reduces to approximately one-half the single mill installation cost. Landfill capital costs are not included in this figure.

Baling facility costs are somewhat more difficult to state because so few operating facilities exist. Total facility costs of one million dollars for a plant capable of processing 900 tons of solid waste per 16-hour day were quoted at the APWA Congress Seminar on High Pressure Baling of Solid Waste in September of 1971. Landfill costs are excluded from this figure. Of course, a complete cost study must be made for any proposed facility.

Note that the equipment needs at the sanitary landfill will be reduced with either milled or baled refuse. One piece of heavy equipment is adequate to handle at least 300 tons of milled refuse per shift and still allow time for working with bulky items, etc., if daily cover is not used. A single fork lift truck is often able to handle up to 60 tons of bales per hour, to which must be added any equipment costs associated with the soil cover.

Operating Costs

The operating costs associated with sanitary landfilling are usually grouped into the categories of labor, equipment, and administration and overhead.

Labor

Wages normally make up about 40 to 50 percent of the total operating costs, or somewhere around $1 per ton of refuse. Labor costs are directly related to the type and amount of equipment, volume of refuse, and characteristics of the site. Larger landfill operations generally obtain greater labor productivity—in terms of man-hours per ton of refuse—due to labor specialization and the use of larger, more specialized equipment. Small, one-man fill operations can frequently abate labor costs of disposal by employing equipment operators on a part-time basis or by assigning landfill workers to other jobs on certain days of the week.

Equipment

The operating costs associated with equipment are rent or depreciation, fuel, and maintenance. The cost of owning and operating a crawler-type tractor has been established in the range of $5 to $15 per hour of operation, averaging slightly over $8 per hour. This does not include labor costs for equipment operators. Equipment costs usually account for 30 to 40 percent of total operating costs, or in the neighborhood of 50 cents per ton of refuse. As with labor, there are economies of scale associated with equipment costs; larger operations obtain greater equipment efficiency through specialization. In addition, the increasing fuel costs will have their effect on total operating costs.

Administration and Overhead

This category of operating costs generally accounts for about 20 percent of total operating costs and includes all the expenses falling outside of labor and equipment costs. The mix of these expenses depends a great deal on the size, type, and location of the landfill project.

Cover Material

This may be considered an additional operating expense (separate from labor and equipment costs) when cover material must be purchased and transported to the landfill site. It may become a seasonal operating cost, because cover material may be imported only during prolonged rainy seasons or stockpiled over winter, etc. Or, cover may become a special operating cost near the completion of a project when on-site sources become depleted. In any event, budgetary allowances should be made for the contingency of importing cover material, if lack of it is indeed a threat to successful operation of the landfill.

The cost of having cover material delivered to a fill site will fall between 75 cents and $3.50 per cubic yard, but this figure will vary directly with the required haul distance.

Total Sanitary Landfill Cost—
Unprocessed Refuse

There are many misconceptions and pitfalls involved in establishing and interpreting the total costs of sanitary landfill operations. Figures ranging from 50 cents to $4 per ton are cited; depending on the hearer's position and bias, figures at either extreme or somewhere in between will be accepted. The reasons for the wide range are complex and must be understood by anyone wishing to use them constructively.

First, there are site differences that influence the cost of cover, special costs due to climate or weather, or special facilities or site preparation needs which must be properly accounted for. The depth of refuse in the completed fill and the original land cost are of considerable importance in determining land costs per ton of refuse disposed.

Second, refuse characteristics vary, ranging from toxic and hazardous wastes, bulky items, tree limbs, etc., to other waste requiring special handling. Even the moisture content of the solid waste can be important, for it is not uncommon for moisture content to vary from 20 to 50 percent on a dry weight basis from location to location. Thus a facility disposing 120 tons per day of refuse at 20 percent water on a dry weight basis would actually be disposing of more refuse than a facility disposing of 150 tons per day at 50 percent water. On this basis alone, the cost per ton could be 20 percent higher for the facility disposing of 120 tons per day.

Third, a factor which is hard to evaluate in monetary value, but which can lead to large differences in sanitary landfilling costs, is the quality of the operation. Adequate equipment plus standby equipment and personnel, good roads in the site, careful policing of the site, careful and prompt application of cover, proper planning and administration, inclusion of adequate site closing costs, etc., are all factors which can affect greatly the total cost of disposal.

Fourth, the size of the operation will influence the cost per ton. All else being equal, large operations are more efficient, providing flexibility in operations, lower site preparation and planning costs per ton, and faster handling of refuse.

Last, there are considerable differences in accounting if certain costs are charged against landfill operations. Land cost is a primary example, for if the land is bought for refuse disposal and used ultimately as a park, it may be appropriate to charge the land costs to the park budget. In small operations in particular, it is not unusual for certain landfill costs to be unknown, or at least not properly charged to the cost of the operation. If street department personnel are used one day a week for administration, or even operation of a portion of the site, for example, their time should be charged accordingly to site operations. Another factor involved in accounting or reporting methods is whether or not a true indication of the tons of refuse handled is available. If scales are not available to weigh all solid wastes coming in to the site, the cost per ton for disposal is nothing more than an estimate subject to great error.

The federal government has attempted to minimize some of the cost reporting problems by publishing a cost accounting system for sanitary landfills. This is reproduced in the Appendix to this report because of its special significance.

Surveys across the country have been taken to determine sanitary landfill costs more accurately. The most widely publicized of these surveys is summarized in Figure 6-1, which indicates that the average cost per ton ranges from approximately $2.50 for small operations to $1 for large operations. The shaded area indicates the range of costs experienced or likely for each size of operation. These figures represent costs in the late 1960s, and should be increased by perhaps 20 to 30 percent to reflect present day costs. Increasing fuel costs should also be taken into account. Note also that these figures must be interpreted carefully, in the light of many of the concerns described earlier in this section.

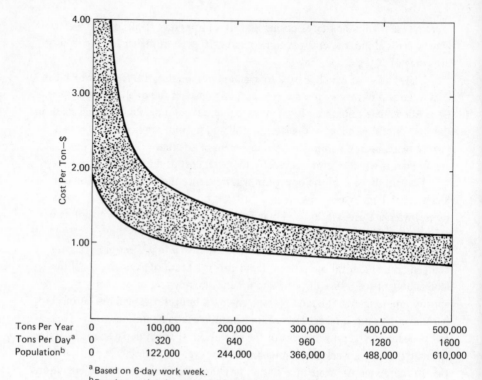

Tons Per Year	0	100,000	200,000	300,000	400,000	500,000
Tons Per Day[a]	0	320	640	960	1280	1600
Population[b]	0	122,000	244,000	366,000	488,000	610,000

[a] Based on 6-day work week.

[b] Based on national average of 4.5 lbs per person per calendar day.

Figure 6-1. Sanitary Landfill Operating Costs. Source: T. Sorg and H.L. Hickman, *Sanitary Landfill Facts,* U.S. Department of Health, Education and Welfare, Bureau of Solid Waste Management, Pubn. No. SW–4ts (1970), p. 23.

Refuse Processing

Refuse processing by milling or baling prior to landfilling is expensive, especially when compared with some of the amazingly low figures widely quoted as the cost per ton for various refuse disposal-reduction methods. The cost of processing must be carefully justified on the basis of decreased hauling costs, decreased disposal costs, quality of operations, and other factors mentioned in Chapters 4 and 5. The outcome of a cost study will involve intangibles such as public acceptance and quality of the operation, and also certain factors dealing with site location and climate. Thus it may be expected that processing will not be justified in many locations even though an operation fifty miles away can properly justify it. Maybe something as simple as an expressway—as opposed to a congested secondary road—between the centroids of refuse generation and the landfill site will determine whether or not to install a processing facility.

At Madison, the total cost of milling, including operating and amortization costs, has been approximately $4 per ton of refuse. To this must be added compaction, hauling and landfilling costs, which would bring the total cost for refuse processing and disposal at the landfill site less than a mile from the processing facility close to $5 per ton, exclusive of land costs. These figures are for a two-mill, two-shift-per-day operation and a one-shift-per-day landfill operation processing approximately 220 tons of refuse per day. The manufacturers rated capacity of the mills totals 23 tons per hour, so it is clear that the mills were not operating at full capacity.

The total cost per ton is only partially a function of the tonnage processed through the facility, for approximately 75 percent of the costs are not related to throughput. Thus, once the facility is obtained and staffed, the additional cost to process a ton of refuse is relatively insignificant. Cost figures must then be interpreted cautiously, remembering that production per day is a major factor influencing cost per ton. Projected costs from Madison, which assume continued operation at 80 percent of the manufacturers' rated capacity of the facility, range from $6.70 per ton for a one mill (15-ton per hour rated capacity) facility operating one shift per day to $2.65 per ton for a four mill installation operating two shifts per day. These figures include the total cost of milling and landfilling the milled refuse at a site within one mile of the processing facility. Installations other than the one at Madison have not published detailed cost breakdowns, but figures in the literature range from $5 to $6 per ton.

Processing costs for a plant baling 900 tons per 16 hour day and operating at 95 percent capacity were estimated at $1.73 per ton at the APWA Congress referred to earlier. This figure includes operating and amortization costs for the baling plant, but excludes transportation and landfilling costs. At St. Paul, the transportation costs average approximately four and one half cents per ton-mile over the entire haul, and the landfill costs are close to $0.80 per ton. Thus, for a haul of twenty-five miles, the total cost would be close to $4 per ton.

As with milling, however, the cost per ton will be a strong function of the rated capacity versus the realized production rate of the facility, and costs realized and projected should be interpreted accordingly. Note that baling costs in excess of $10 per ton have been experienced during initial operations; it is unlikely that these costs will be typical of long-term, large-scale operations.

7 Public Health Considerations

The main reason for the existence of sanitary landfilling, as opposed to open dumping of refuse, is that it offers superior public health and safety and presents less threat to the surrounding environment. A sanitary landfill must live up to the standards outlined in previous chapters, or it is, in effect, little better than open dumping.

In order to ensure that a sanitary landfill does not become a threat to public health and safety during its construction and after its completion, many states have begun requiring inspection of landfill operations by the state department of health. Even if such inspection is not mandatory under state law, it is recommended that the community that disposes of its solid wastes by landfilling seeks the assistance of its state health department to make certain they are not "polluting their own backyard."

The three main areas of concern with regard to public health and safety are: (1) potential spread of disease from contaminated refuse, (2) potential contamination of ground and/or surface water, and (3) production of gases which may be hazardous to the surrounding environment.

Vector Control

The daily earth cover placed over compacted refuse will greatly reduce the presence of birds, insects and vermin. The cover reduces the attractiveness of the area to vectors and limits the amount of burrowing into fill cells and feeding on organic wastes. A minimum of six inches of compacted earth will also prevent flies trapped in the refuse from escaping to the surface, depending on the soil.

For open dumps subsequently converted to sanitary landfills, such as Kenilworth in the District of Columbia, it is usually necessary to conduct extensive rodent and insect extermination prior to compacting and covering the previously dumped refuse. Such extermination activities should be conducted by properly equipped professionals. When pre-fill extermination involves toxic chemicals, special efforts must be made to incorporate all of the treated refuse into the fill as quickly as possible so that rain will not create toxic run-off waters.

Problems created by vectors are commonly associated with solid waste by the public. A sanitary landfill virtually eliminates vector problems by compacting the solid waste and promptly covering it with earth. If there is proper extermination prior to site operation and the operating requirements of a sanitary landfill are followed, there should be little if any vector problem. The one area of practical concern is the possible presence of birds around any area, including even a sanitary landfill, where solid waste is found, and thus sites should not be located adjoining airports. The prompt use of cover will minimize but not necessarily eliminate the presence of birds.

Water Pollution

General Observations

Water pollution can result even from an otherwise properly operated sanitary landfill and can in turn pollute soil. Even if a fill project has been well planned, designed, and constructed, some water pollution will occur in the immediate vicinity of the fill in a high rainfall climate. It is inevitable that a portion of the rainwater will seep through the cover material and into the refuse. This water, plus the original water content of the refuse, will tend to move from within the boundaries of the landfill and eventually pollute surrounding soil. The polluted effluent from a solid waste landfill is called leachate, and when leachate seeps into underground aquifers (water-bearing strata), the threat of water pollution becomes relevant for areas outside the immediate vicinity of the fill site.

The seriousness of the pollution of soil and ground water surrounding a landfill has recently become an area of increasing concern, partly because only within the past decade has any extensive research been conducted in the area of landfill leaching. On one side well meaning but often uninformed citizens use emotional arguments based on hearsay evidence of gross contamination of water resources by solid waste disposal. Cases where people have opposed new landfill sites at public hearings because of concern about pollution of their well waters, even though the site is up to fifteen miles away, are altogether too common. On the other hand, landfill owners and operators, and others who stand to benefit from a new site, sometimes overlook or minimize the danger, even though there is a definite possibility of ground or surface water quality degradation.

Recent evidence seems to suggest that there are very few documented cases where water pollution has resulted from solid waste disposal. The report by

Zanoni, which records the results of a careful search of the literature and a survey of states, states:

> After some review of the literature on sanitary landfills, one point becomes clear almost immediately: there are very few case histories of serious or even troublesome contamination of ground water which are directly attributable to the leachate from sanitary landfills. There may, of course, have been unpublished instances or unknown cases—but the writer has not learned of a single person who has died as the result of ground water being contaminated by a landfill. Considering the number of landfills past and present and the amount and variety of solid wastes generated in our modern technical society, this is a remarkable situation.[1]

He goes on to state that, "the above situation attests to the almost miraculous capability of most soils to attenuate the leachates generated from sanitary landfills," and presumably this would include open dumps as well.

Cases of water pollution from landfills usually involve surface waters, where the ability of the soil to purify waters is not involved, and where a direct link exists between the landfill and surface water. Those cases involving ground water are cases of obvious negligence, such as when a landfill is placed right in an aquifer with a well immediately downstream, or when ground water quality is being monitored right in or very near the solid waste. It should be noted that if ground water contamination is observable, it will involve contaminants generally found in natural ground waters anyway. This is because the soil will remove practically anything else. Thus, the pollutants originating from solid waste disposal that are most likely to be found in ground water are not classified as health hazards—with the possible exception of nitrates, for which the United States Public Health Service (USPHS) drinking water standard recommends a maximum of 10ppm.

Background

In studies conducted in California by the State Water Pollution Control Board, it was demonstrated that within a year after initiation of one test landfill, continuous leaching of one acre-foot of fill would result in the leaching of 1.5 tons of sodium and potassium, 1.0 ton of calcium and magnesium, 0.91 ton of chloride, 0.23 ton of sulfate, and 3.9 tons of bicarbonates. Thus it is apparent that substantial amounts of pollutants may be removed from solid waste landfills and transported to the surrounding soil or water systems.

Subsurface water usually occurs in either a zone of aeration or a zone of saturation. A zone of aeration is a condition where the pores of soil and rock are filled with both air and water; in a zone of saturation, pores are entirely filled with water. The upper surface of a zone of saturation is frequently referred to as a water table. Some water is held in the zone of aeration by the attractive force that exists between soil particles and water, but water in excess of the amount that can be held by these forces moves downward under the influence of gravity. The amount of water in the zone of aeration depends on precipitation, vegetation cover, depth of the water table, and soil conditions. The ease with which water moves through a zone of saturation depends on the permeability of the soil. Sand and gravel strata allow more rapid water movement than those of silt or clay. Joints or fractures in rigid rock structures (such as limestone, shale, sandstone, and granite) also usually provide for easy movement of underground waters.

Water from precipitation or leachate from a landfill tend to move downward through the zone of aeration, but once they reach the zone of saturation, they move with the ground water. This ground water movement can be either local (e.g., heading for the nearest discharge point into a swamp, pond, or surface stream) or regional (usually deeper than and often running underneath local discharge points and toward major points of discharge—rivers, lakes, oceans). Fine-grained soil with small pores acts as an effective filter for bacteria, but such filtration is not effective for removal of dissolved chemicals. When landfill leachate has direct access to water-carrying fissures in rock, little or no removal of either biological or chemical contaminants occurs through filtration. Therefore, special caution must be exercised when constructing landfill in abandoned quarries and strip mines to make sure refuse does not come in contact with exposed aquifers. Note that special caution should also be exercised at landfills located where leachates have direct access to surface waters.

> The practice of requiring at least 30 feet of relatively impermeable material between the base of a landfill and any aquifer has been followed for many years in Illinois . . . recent research . . . suggests that it is a reasonable though perhaps conservative requirement.[2]

However, in California the practice has been to require from three to ten feet of soil—depending on the soil type—between the base of the landfill and the ground water table.

Experimental tests on landfill leaching conducted by South Dakota State University have indicated that the parameters which best serve to describe chemical pollution of ground waters from landfill leaching are: pH, specific conductance, total hardness, calcium hardness, alkalinity, chloride and nitrate content. In these experiments, tests were also conducted to determine the presence of sodium, potassium, and iron. It was determined that chemical pollutants were removed from underground water 1,200 feet downstream from the fill. This was established by drilling thirty-one test wells of varying depth and distance from the fill area. Samples drawn adjacent to the downstream side of the fill indicated chloride content fifty times as great as that of upstream control wells. Hardness and alkalinity were nearly twice as great in the downstream well samples; sodium and potassium content was over twenty times as great; and iron content was sixty times as great. In tests of statistical significance for the increase in concentration of all the samples drawn from the downstream wells, the various test parameters ranked in the following order as most likely to vary from control well samples: chlorides, sodium, nitrates, specific conductance, hardness, alkalinity, and pH. Therefore, a landfill operator desiring an expedient test of whether his operation is contaminating underground waters may wish to test for chlorides, sodium, and nitrate content of a well sample drawn 500 to 1,000 feet downstream from the fill site. In order for these tests to be meaningful, however, a control sample drawn 50 to 100 feet upstream from the fill must be tested to determine the chemical content of ground waters entering the region.

Microorganisms have been found in significant numbers in leachate from sanitary landfills. The most commonly found are various strains of mesophilic aerobes, thermophilic aerobes, anaerobes, actinomycetes, and fungi. Concentration of microorganisms in leachate may be expected to peak within nine to eighteen months after fill operations commence, but substantial numbers of microorganisms have been recorded up to five years after fill completion. The health significance of the microorganism content is that pathogens buried with refuse can be transported out of the fill into underground water supplies. However, as indicated previously, biological contaminants are usually filtered by surrounding soil more quickly than chemical contaminants, and for this reason it is very unlikely that water in wells and streams more than a few hundred feet from the fill site will be affected. Furthermore, because of the high temperatures developed in an actively decomposing sanitary landfill, pathogens are generally killed, making pathogen contamination of ground waters unlikely.

Leachate Control

There are several available methods of controling or minimizing the effect of leachate from sanitary landfills. It should first be cautioned that these methods should not be applied indiscriminantly in an attempt to make all sites "environmentally acceptable." Rather, the possibility of leachate contamination of ground or surface waters should be competently assessed, and the most appropriate method of countering it should be selected for each site. Water contamination is of concern when the solid waste is placed near surface or ground waters, when there is relatively unhindered flow from the solid waste to the surface or ground waters, and when the distance from the landfill to a water user is short. Thus, placement of refuse in standing water and the location of sites over coarse soils or fractured bedrock or near wells or surface bodies of water should be avoided. When such sites are used, special precautions should be taken to minimize water pollution, and these are the methods outlined herein.

Diversion of surface and/or ground water away from solid waste is the most sensible method of leachate control, in the view of many engineers and geologists. The concept is to minimize or control the *production* of leachate, rather than attempting to deal with the leachate produced. Reduction of the inflow of ground and surface water to the refuse will reduce the production of leachate.

Ground water flow into the refuse means that the refuse extends below the water table. Normally such a site will be avoided, but in cases where use of the site is of overriding importance, one can separate the ground water from the solid waste by a liner or divert the ground water away from the solid waste. A liner provides a watertight seal, perhaps a plastic membrane or clay, to keep the ground water from flowing into the solid waste. Liners will be discussed further later in this section. Diversion may involve a trench, upstream from the site, which is pumped out to lower the water table; or it may involve a series of wells or trenches around the site which have been pumped out to lower the water table. The latter system is in use at the Mountain View sanitary landfill south of San Francisco. The site is in wetlands adjacent to San Francisco Bay and would be unacceptable—without special precautions—for a sanitary landfill because of the possibility of water contamination. The site is divided into relatively small areas, each of which is surrounded by trenches dug well below the deepest level to be filled with solid waste. These trenches are pumped to lower the water level below that of the solid waste so that there is no direct contact. Pumping must continue until the landfill is stabilized and there is no further leachate production.

Diversion of surface waters away from a sanitary landfill has been considered good practice for many years, but has normally not been used to control leachate production. Diversion of surface water with trenches or drainage ditches uphill of and around a site reduces mud and standing water on the site. By diverting water away from the site, the amount of infiltration and therefore the production of leachate is reduced.

The use of diversion for control of leachate is considered in a paper by Salvato, Wilkie, and Mead on the prevention and control of leachate from sanitary landfills.[3] The water budget at the surface of a landfill indicates that incident precipitation can leave the surface as runoff, infiltrate into the landfill surface, evaporate, or transpire back into the atmosphere. Water that infiltrates into the surface may then percolate downward through the underlying material, or transpire through vegetation to the atmosphere. To minimize leachate production, runoff should be maximized by using a tight, smooth, and sloped soil cover; infiltration should be minimized by using a tight soil cover; percolation should be minimized by using a tight impermeable material under the surface; and transpiration should be maximized by using plants able to transpire large volumes of water. Of course, not all of these parameters are equally important in a given landfill, for a high runoff coefficient may override the importance of the infiltration or percolation rates of the particular soil in question. An indication of the range and relative importance of some of these rates of water flow is abstracted from the original paper in the tables that follow.

The final method to be discussed for limiting the amount of leachate produced is applicable in sites through which streams are running. It is common in such cases to provide a culvert or pipe to channel the stream through or under the landfill without contacting the solid waste.

Aside from minimizing leachate production by the methods discussed above, it is possible to regulate or control the leachate that is formed in a sanitary landfill to minimize adverse effects of the leachate on ground or surface waters. One widely publicized method of control is to line the site with a barrier impermeable to leachate flow. Clay liners or berms have been used at

Table 7-1
Runoff

Type of Surface	Maximum Amount of Runoff per Acre,[a] (gal.)		
	Open Sandy Loam	Clay or Silt Loam	Tight Clay
Meadow, with cover crop:			
flat (0-5% slope)	16,355	49,005	65,340
rolling (5-10%)	26,136	58,806	89,842
hilly (10-30%)	35,937	68,607	98,010
Raw soil surface:			
flat (0-5%)	49,005	81,675	98,010
rolling (5-10%)	65,340	98,010	114,345
hilly (10-30%)	84,942	117,612	133,947

[a] Amount of runoff = CI, where C = coefficient of runoff, I = intensity of rain, assumed at one inch per hour for six hours.

Table 7-2
Percolation

Cover Material	Possible Quantities of Water Reaching Refuse[a]
Uniform coarse sand	434,000,000 gal/acre
Uniform fine sand	4,340,000
Graded silty sand and gravel	434,000
Silty sand	97,800
Sandy clay	5,200
Clay	98

[a]In 24-hour period. Assumptions are cover material is saturated and uniform, there is no resistance to flow below the cover material, a covering of water is present over the cover, and Darcy's Law is applicable.

Table 7-3
Transpiration

Plants	Appropriate Seasonal Consumption of Water by Plants	
	Inches	Thousands of Gallons per Acre
Coniferous Trees	4–9	109–245
Deciduous Trees	7–10	190–270
Potatoes	7–11	190–300
Wheat	20–22	540–600
Grapes	6 up	163 up
Alfalfa and Clover	2.5 up	68 up
Corn	20–75	540–2,040
Oats	28–40	760–1,090
Meadow Grass	22–60	600–1,630
Lucerne Grass	26–55	706–1,500
Rice	60–200	1,630–5,400

many sites for this purpose. The DuPage County sanitary landfill hill, for example, was constructed over a natural clay deposit, and berms of clay were placed around the filling area at a depth sufficient to extend into the natural clay layer, in effect creating a liner. This site will be discussed in more detail near the end of the report. Bituminous fiber, rubber, and several types of plastic film are among the other liners available.

The sanitary landfill operated by the Warner Company, a Philadelphia-based sand and gravel company in Bucks County, Pennsylvania, has a membrane barrier to leachate flow covering its base. The membrane is protected by putting a cushion of sand over it before refuse is placed, and by appropriate grading and/or filling prior to placement. The design of the site is especially noteworthy, for it will produce a 182-foot recreational hill of refuse and cover material for winter sports, etc. A site at Romeo, Michigan, operated by the Predmore Development Company also contains a plastic membrane liner, in this case supplied by Staff Industries, Inc.

For liners under a landfill to be successful, they must not leak. This is difficult to achieve in practice, but apparently it can be done. If liners do not leak, they must contain leachate. Unfortunately, this alone is insufficient for extended periods of time, and there are altogether too many sites with substantial volumes of leachate ponded on the site. Cases are known where well over ten feet of leachate has ponded on sites of many acres, because only leachate collection was considered in the site design. The potential damage this represents is considerable, especially when the extremely polluted nature of leachate is considered. Therefore, implicit in the proper use of liners is not only leachate collection, but controlled discharge, possibly including treatment. Leachate treatment is included, for example, in the Bucks County landfill. An EPA-funded project at the University of Wisconsin has shown that leachate is treatable by biological and chemical means, but that even after treatment the load of certain contaminants as chlorides, sodium, potassium, and residual organics may be sufficient to be of concern. Two other EPA projects, a research project at the Georgia Institute of Technology and a demonstration project at Sonoma County, California, are examining the recycling of leachate through landfills to promote stabilization of solid waste. The results from these studies are not yet available, but the concept is interesting and promising.

There are some natural site conditions which are readily adaptable to leachate collection. These conditions involve a site with an initial slope, such as on the side of a hill or in a gully. The Kent landfill south of Seattle, Washington, is such a site. Here leachate is collected by a series of trenches dug into the original soil surface across the inclined faces. Leachate tends to work downward through the solid waste until it meets the underlying soil. It then flows down the original soil slope until it meets an intercepting drainage ditch. The ditches bring the leachate to a common point for treatment.

The use of a site adjacent to surface water bodies frequently involves construction of an impermeable berm between the landfill and the water body. Often the terrain of such a site is such as to bring the leachate to a common area for collection or treatment.

Another method of collecting leachate is to determine the motion of ground water containing leachate and to build an intercepting trench, possibly backfilled with coarse gravel to collect the contaminated water. Care must be taken to determine carefully the hydrogeology of a site to be sure that the trench's placement and dimensions are such as to insure collection of all of the leachate. At Madison, Wisconsin, the Olin Avenue landfill is located adjacent to a stream. Hydrogeological investigations indicated that ground water was moving across the site into the stream. A gravel filled trench was constructed between the landfill and the stream, and the collected water is put into a sewer for treatment at the sewage treatment plant.

At Brookings, South Dakota, the direction and depth of ground water flow across the city landfill is clearly defined, so a trench was dug to intercept ground water containing leachate immediately downstream from the site.[4] The dimensions of the trench are 1,100 feet long by fifty feet wide and fifteen feet deep. By diluting the leachate with relatively uncontaminated ground water, treating it during holding in the trench, and/or lowering the ground water level away from the refuse, the trench appears to have halted the trend toward rising contaminant levels in downstream well waters. It is further reported that ground water in the immediate vicinity of the site nearly meets USPHS drinking water standards. The water is of questionable value for this purpose, but cannot be ruled out altogether. Downstream from the trench, the water quality was acceptable for domestic water supply.

Decomposition and Gas Production

In contrast to leachate, which has been the cause of few if any cases of documented damage in sanitary landfills, gas from decomposing refuse in landfills has caused documented property damage and led to injuries and death. Gas is therefore of major concern to most engineers and planners and requires careful evaluation and possibly the application of control measures for each site.

The decomposition of refuse in a sanitary landfill depends on several different factors, such as permeability of earth cover, depth of fill, rainfall, degree of refuse compaction, and the refuse moisture content and putrescibility. Microorganisms within surrounding soil and the refuse itself are responsible for decomposition. As a rule, decomposition takes place quickly until the atmospheric oxygen supply is reduced to basically zero. The greatest temperature elevations occur during this period, which generally lasts no more than a few weeks. When the atmospheric oxygen level drops to zero, anaerobic decomposition begins, and the temperature drops. After many successive breakdowns through feeding and growth of the microorganisms and the action of their enzymes, the end products are gases and a humus composition. The major gaseous endproducts of refuse decomposition are carbon dioxide (CO_2) and water for aerobic decomposition, and CO_2, methane (CH_4), ammonia (NH_3) and hydrogen sulfide (H_2S) under anaerobic conditions.

Most fill sites will actively decompose for many years. The length of the period of decomposition will depend on such factors as refuse composition, depth of fill, and local climatic conditions. Fills in Louisiana and Virginia have reportedly "decomposed" in three to four years; other fills have remained active for over twenty years. Glass, plastic, and aluminum refuse in a fill will remain relatively unaltered for twenty years, and depending on the moisture content, paper products may also remain intact for that length of time. Steel, most woods, cloth, and other organic wastes will be partially to totally decomposed within twenty years.

Table 7–4 indicates in more detail what happens to various components of municipal refuse in landfills.

Methane and carbon dioxide are the principal gases produced in a landfill, and various sources indicate that the production of carbon dioxide reaches a peak within six months to two years and that the production of methane will tend to increase, although at a slower rate, steadily over several years. Gas generation estimates have been given by Coe as 0.035 and 0.21 cubic feet of gas per pound of dry and saturated refuse respectively.[5] Concentrations of carbon dioxide in a fill may be as high as 90 percent of total gas volume, while concentrations of methane have been recorded as high as 65 percent.[6] Production of CH_4 increases as the moisture content of the buried refuse increases. Conversely, as the amount of water increases, the CO_2 levels in the gases go down, and production therefore seems to be reduced. This may not represent a true reduction in CO_2 production, however, because the high solubility of CO_2 in water is likely providing accelerated removal of gaseous CO_2. A result of its solubility is that carbon dioxide is the only gas which appears in significant quantities in the soil and water surrounding a sanitary landfill, usually resulting in the production of carbonates and bicarbonates. The methane and other gases produced tend to rise through porous cover material and escape into the atmosphere.

Experiments conducted at landfill sites near Oakland, California, have indicated that carbon dioxide tends to move out of the fill cell through surrounding soil at a vertical velocity of about 0.2 feet per day, and that the horizontal velocity is about the same. It can be expected that sizeable concentrations of CO_2 will remain in contact with soil immediately adjacent to the fill for many years; however, it is during the first two years of a fill's life that excessive CO_2 concentrations in the surrounding ground water can be expected to occur. The amount of gas which will actually be picked up by ground water depends on a variety of factors, such as depth of the water table, thickness of the aquifer, velocity of ground water, and soil permeability.

It is the limitation of free movement of CH_4 to the atmosphere that is of special concern. CH_4 at concentrations ranging from approximately five to fifteen percent in air is explosive, so somewhere in the system between the thirty

Table 7-4

Environmental Rate and Conversion Products of Domestic Wastes in Sanitary Landfill

Metals (Approx. 7% of total)

1. Remain in landfill as inert or relatively inert compounds unless fill eroded by surface runoff: Uncombined metals; Al_2O; $Al(OH)_3$; FeO; FeO_2; $Fe(OH)_2$; $Fe(OH)_3$; CrO_2; Cr_2O_3; $Cr(OH)_2$; Cu_2O; $Cu(OH)_2$; SnO; ZnO; $Zn(OH)_2$; V_2O; Be; BeO; NiO; Ni_2O_3; $Ni(OH)_2$; CdO; Cd_2O; $Cd(OH)_2$; PbO; Pb_2O_3; Pb_3O_4; Pb_2O; Se; SeO_2; Hg; HgO; TiO_2; MgO; $Mg(OH)_2$; $Ca(OH)_2$; $CaCO_3$.

2. May leach to groundwater: Ca and Mg sulfates; Fe, Ca, and Mg bicarbonates; CO_2; also oxides of Sn, Zn, Cu in acid water.

3. Escape to atmosphere: essentially none.

Cloth-Natural & Synthetic (4% of total)

1. Remain in landfill as inert or relatively inert materials:

a. Materials fabricated of synthetic fabrics.

b. Oxidized and reduced minerals in "natural" fibers.

2. Remain in landfill by incorporation into microbial protoplasm: NH_4^+; reduced sulfur compounds: C; P; K.

3. May leach out to groundwater: CO_2; aldehydes; ketones; organic acids; sulfates; phosphates; NH_4^+; NO_2^-; NO_3^-.

4. Escape to atmosphere: CO_2; CH_4; volatile short-chain fatty acids; N_2; NH_3; H_2S; mercaptans.

Plastics (1.9% of total)

Essentially inert in landfill.

Leather (1% of total)

1. May leach out to groundwater: CO_2; aldehydes; ketones; organic acids; sulfates; phosphates; NH_4^+; NO_2^-; NO_3^-.

2. Remain in landfill by incorporation into microbial protoplasm: NH_4^+; reduced sulfur compounds; C; P; K.

3. May escape into atmosphere: CO_2; CH_4; volatile short-chain fatty acids; N_2; NH_3; H_2S; mercaptans.

Rubber-Natural & Synthetic (1.1% of total)

1. Synthetic rubber is essentially inert.

2. Natural rubber breaks down extremely slowly.

Glass (8% of total)

Inert in landfill.

Wood (2% of total)

1. May leach to groundwater: CO_2; aldehydes; ketones; organic acids; phenol; NH_4^+; NO_2^-; NO_3^-.

2. Remain in landfill through incorporation into microbial protoplasm: NH_4^+; C; P; K.

3. May escape into atmosphere: CO_2; CH_4; volatile short-chain fatty acids; N_2; NH_3.

Garbage (15.5% of total)

1. Possible leachates to groundwater; See *WOOD* (above); plus sulfates, phosphates, and carbonates.

2. See Item 2, WOOD (above).

3. May escape into atmosphere: CO_2; CH_4; volatile short-chain fatty acids; H_2S; mercaptans; N_2; NH_3.

Paper (51.5% of total)

1. May leach to groundwater: CO_2; aldehydes; organic acids; phenol; NH_4^+; NO_2^-; NO_3^-.

2. Remain in landfill through incorporation into microbial protoplasm: NH_4^+; C; P; S.

3. May escape into atmosphere: CO_2; volatile short-chain fatty acids; H_2S; mercaptans; N_2; NH_3.

Unclassified (5% of total)

1. Relatively inert.

2. Ashes in fill may leach soluble minerals to groundwater reducing its chemical quality.

Source: Sanitary Engineering Research Laboratory, *Comprehensive Studies of Solid Wastes Management, Second Annual Report* (Berkeley: University of California Press, January 1969).

percent or higher concentration of CH_4 in the landfill and the nil concentration of CH_4 in the atmosphere an explosive gas mixture will be obtained. It is the purpose of proper design to insure that such explosive mixtures are not formed where they can cause harm or damage, such as under or within buildings or other confined spaces. Methods of control range from providing barriers to gas flow—to restrict the movement of gases and confine their presence to remote or otherwise acceptable locations—to promoting free venting of gases to the atmosphere.

Gas Control

The control of gases produced in a landfill during decomposition of solid wastes should be considered in every well-designed landfill. In most cases no special action will need to be taken for gas control, but the potential danger represented by improper venting of gases is of sufficient importance to warrant concern. Of special importance in evaluating the ability of a landfill to vent gases are the properties of the cover soil. One of the problems associated with the use of a tight cover material, as for control of the amount of leachate formed, is that it not only minimizes percolation of water downward, but also these same impermeable characteristics will limit the free venting of gases to the atmosphere.

Gas flow through soils is characterized by two main mechanisms: diffusion and convection. Diffusion may be described by the diffusion coefficient D, which is a function of the soil type and the pressure gradient. Knowing D, the rate of flow can be calculated by Fick's Law:

$$F = DA \; \frac{\partial C}{\partial X}$$

where F = flow in volume gas per second.

D = diffusion coefficient in area per second.

A = cross sectional area through which the gas flows.

$\dfrac{\partial C}{\partial X}$ = concentration gradient with respect to distance, through the soil volume in question, perpendicular to A.

Table 7-5 is adapted from the report entitled "Development of Construction and Use Criteria for Sanitary Landfills" and indicates the effect of the type of soil on the flow of methane by diffusion as determined in test columns of air dry soils.[7] Note that if the soil is wet D may decrease, in some cases by a factor

Table 7-5
Methane Diffusion

Soil Type	D $(Cm^2/sec.)$ at Press. drop ¼ in. H_2O	CH_4 flow[a] (ft^3/sec) at various pressure drops		
		$\Delta P=$¼ in. H_2O	$\Delta P=4$ in. H_2O	$\Delta P=12$ in. H_2O
Gravel	1.34	12.96	364.40	566
Sand	0.0575	0.57	5.60	47
Sandy Silt	0.0317	0.31	1.47	23
Silty Clay	0.026	0.25	0.85	0.90
Clay	nil	nil	nil	1.2×10^{-4}

[a] Assumes a soil barrier 100,000 square feet in area and five feet thick, where the CH_4 concentration reduces from 50 percent to 5 percent across the foil barrier.

Table 7-6
Methane Convection

Soil Type	CH_4 flow[a] (ft^3/sec) at various pressure drops		
	$P=$¼ in. H_2O	$P=4$ in. H_2O	$P=12$ in. H_2O
Gravel	57.0	2515.0	4760.0
Sand	1.0	335.0	1106.0
Sandy Clay	0.67	27.65	96.5
Silty Clay	1.0	5.70	22.4
Clay	nil	nil	1.34

[a] Assumes a soil barrier 100,000 square feet in area and five feet thick, where the gas is 50 percent methane.

of five or so, and that the results are dependent on the degree of compaction of the soil, on temperature, etc.

Convection occurs in addition to diffusion. Convection is the free flow of gas under the action of a pressure gradient. Table 7-6, adopted from the same report as the previous table, provides information derived experimentally on the effect of various air dry soils on the flow of methane by convection. Note that if the soils are wet the flow may be reduced by a factor of twenty or more, and that the results are dependent on the degree of compaction of the soil, temperature, etc.

It is apparent from the above tables that the total rate of gas flow is a strong function of the type of soil, and in some cases may be very high. In Los Angeles, California, a CH_4 concentration of 10 percent (5 to 15 percent CH_4 is considered explosive) was measured 600 to 700 feet from a landfill, illustrating the long distance gases may travel under certain conditions.

Control Measures

Gas control methods are designed to provide free venting of landfill gases in a controlled and acceptable manner. The methods may be divided into two general categories, differing only in whether the soils surrounding the landfill in question are permeable to gas flow or not. If the surrounding soil is impermeable, it is necessary to provide venting on the landfill site itself. If the surrounding soil is permeable, there is a possibility that landfill gases can travel long distances through the soil and cause damage at seemingly great distances. In this case, it is necessary to contain or cut off gas flow before it reaches an area which could be damaged.

If a landfill is located in impermeable soils, gases will not travel laterally away from the site. If an impermeable cover material is used, as is frequent in such situations since cover material is normally obtained on site, the gas pressure will build up until the easiest path to the atmosphere is finally found. Relatively large amounts of gases will then pass through the break(s), probably in an uncontrolled (and possibly dangerous) fashion.

The most logical gas control method is to use a permeable cover material and to avoid paving or building on the surface for many years. Unfortunately, such material may not be readily available, and the increased infiltration of water may be undesirable. If impermeable cover is used, special venting devices must be provided. A network of perforated pipes inserted in the landfill to penetrate the various refuse cells found in the landfill may be provided. The pipes may vent freely to the atmosphere, but they should be long enough to be well out of the reach of passers-by. The preferred vent is a burner of some sort, such as the "TIKI" burner, which safely burns the gases and removes any potential dangers associated with free venting. The Tiki burner is a simple device extending perhaps ten feet into the air. It is equipped with a windshield, and may be manually or automatically ignited with a pilot flame. A major drawback with these devices is the fact that without a network of perforated feeder pipes, each burner vents only a small portion of the landfill. Depending on the success in collecting landfill gases and on the rate of decomposition of the refuse, there may be sufficient gas to support a continuous and sizeable flame. The DuPage County hill described later in this report and a landfill adjacent to the University of Washington in Seattle are just two of many examples of the use of venting devices.

An alternative, or perhaps complementary, device to Tiki burners is a network of trenches around and possibly through the fill. The trenches should be located to catch any gas buildup, and filled with gravel to provide easy venting. A gravel liner under and around a site is another way of providing free venting.

In a large site, it is likely that venting with pipes will also be necessary to take care of portions of the landfill not near a gravel layer or trench.

If a landfill is in a soil permeable to gas flow, vents or gas flow barriers may be used. A trench filled with gravel is very effective either around a site, or at least between the landfill and the area to be protected. At a landfill near Torrance, California, a trench was used to intercept gas flow between a 55-acre, 100-foot-deep sanitary landfill and a house 115 feet from the landfill. Gas compositions of 42 percent CH_4 and 33 percent CO_2 were measured in the yard adjacent to the house. A trench twenty feet deep by thirty inches wide by 1,050 feet long, backfilled with gravel, reduced the gas concentrations in the yard to zero percent CH_4 and 2.2 percent CO_2. Ideally, the depth of the trench should be greater than the depth of the refuse, but certainly it should be of sufficient depth to intersect any permeable gas-carrying layers.

A series of gravel filled holes may be dug in lieu of a trench. The potential problem with holes is that they may miss much of the gas flow if placed too far apart, yet it is expensive to dig holes close together. To reduce the number of holes, it may be worthwhile to pump the gases out of every other hole, creating gas flow through the soil from the open holes to the holes being pumped out. This flow of gases will help flush out the area between holes, in effect creating a trench. This concept was applied in Los Angeles, California, using five holes, sixty feet deep and thirty inches in diamater, spaced forty feet apart.

Another method of controlling gases in permeable soils is to construct a barrier to gas flow. Whenever barriers are used, it is, of course, necessary to provide venting of gases trapped by the barrier, as with perforated pipes, etc. Compacted tight soil (fine silt or clay) forms a nearly impermeable barrier to gas flow and may be used as a liner around a landfill site. If the walls of a site to be provided with a liner are inclined fairly steeply, a 10-feet thick liner will normally have to be used just to accommodate the heavy equipment needed for building the liner. A liner constructed of tight soils must be kept wet, otherwise the soil becomes more permeable and, more importantly, subject to cracking.

Gunite is a pneumatically applied mortar, which may be placed around a landfill site. It is particularly useful on slopes. It should be applied carefully to avoid cracks. Bituminous-fiber liners have been used, as in one case near Torrance, California, where the liner protected a building located over a sanitary landfill. Four inches of gravel imbedded with perforated pipe for venting was placed to carry gas on the refuse side of the liner; this followed by three inches of sand, the liner, and then a 6-inch layer of pea gravel to protect the liner.

Plastic sheets are difficult to use because they may rupture. Furthermore, the choice of plastic is important because some plastics are fairly permeable to gas flow. PVC film ten mils thick was found in special tests to reduce CO_2 concentrations by 44 percent; whereas with an 8-mil thick polyethylene film the reduction was 80 percent.

In summary, gas concentrations reaching explosive proportions are found in virtually all sanitary landfills, and the design of a landfill must include the determination of where these gases will go. Any construction over a landfill must be done with a great deal of caution, because such construction represents a barrier to gas flow and thus a buildup in concentration. Explosions have been touched off by heat in enclosed areas where landfill gases reached explosive concentration. Buildings can provide such areas if appropriate precautions are not taken.

Notes

1. A.E. Zanoni, "Ground Water Pollution from Sanitary Landfills and Refuse Dump Grounds—A Critical Review," Department of Natural Resources, Research Report 69 (Madison, Wisconsin, 1971), p. 12.
2. K. Cartwright and F. Sherman, *Evaluating Sanitary Landfill Sites in Illinois.* Environmental Geology Notes no. 27 (Urbana, Ill.: Illinois State Geological Survey, August 1969), p. 23.
3. J.A. Salvato, W.G. Wilkie, and B.E. Mead, "Sanitary Landfill—Leaching Prevention and Control," *Journal of the Water Pollution Control Federation*, 43(10): 2084, October 1971.
4. J.R. Anderson and J.N. Dornbush, "Intercepting Trench at Landfill Renders Leachate into Drinkable Source of Water," *Solid Waste Management/RRJ*, 15(1): 60, January 1972.
5. J.J. Coe, "Effect of Solid Waste Disposal on Ground Water Quality," *Journal of the American Water Works Association*, 62(12): 776, December 1970.
6. A.A. Fungaroli, personal communication with R.M. Vancil, dated May 1, 1973.
7. County of Los Angeles, Department of County Engineer, *Development of Construction and Use Criteria for Sanitary Landfills*, interim report to U.S. Department of Health, Education and Welfare, Bureau of Solid Waste Management, (Cincinnati, 1969), p. 48.

8

Alternative Land Uses and the Future of Sanitary Landfilling

Of the nearly 115 million tons of municipal solid wastes generated in this country annually, as little as five to ten million tons may find its way into what could be classified as a true sanitary landfill, while as much as 110 million tons are deposited at land disposal sites which will be unfit for future land use without extensive renovation.

The fact that land which has served as a refuse disposal site can become an asset to the community and be used for a variety of purposes should provide additional impetus for the continued conversion of open dumps and open burning sites to sanitary landfills. This is perhaps the most undersold and neglected positive aspect of sanitary landfills. The imagination and goodwill of the public might be prompted by emphasizing this positive aspect of nearly all well-conceived sanitary landfill projects.

Alternative Land Uses

The most popular use of completed landfill sites near populated areas has been for recreational purposes. This is understandable because it facilitates public acceptance of the project, and because subsequent settling and gas production present less threat to life and property since few permanent structures are erected over the fill. Parks, playgrounds, ball fields, and golf courses are suitable final land uses for most fill projects, presenting few operating difficulties and requiring little additional advance planning.

More exotic recreational facilities for which landfill disposal sites have been used in recent years include a variety of high-rise projects destined to become ski slopes, toboggan runs, amphitheaters, and soap-box derby tracks. It usually becomes necessary to stack refuse as high as 100 feet or more above the original elevation to carry out such projects. As might be expected, this creates some technical difficulties above and beyond those of traditional 10- to 20-feet deep fills. Most operating difficulties can be overcome by incorporating more earthern cover material into the fill than one normally expects to find in an ordinary trench or area method fill project, or possibly by processing the refuse before landfilling. Unusual landfill designs will generally increase the overall disposal cost per ton of refuse; however, there is usually adequate community support for such a project, so that the additional operating expense of high-rise refuse disposal is willingly absorbed. Another relevant consideration when planning an elevated fill is that the large amounts of refuse and fill material required to create a hill of significant size may necessitate a five to ten year project, during which time enthusiasm for the finished project may have diminished. There are also special operating problems created by slopes of 30 to 45 degrees on the sides of a mound, so that periodic regrading and prompt seeding of a grass cover may be necessary. The DuPage County landfill described later in this section provides a case history of such a recreational hill.

In the event that a parking lot is planned for a completed landfill disposal site, a flexible paving with a bituminous binder is recommended to prevent cracking from the inevitable settling. For landfill projects accepting incinerator residue or solid demolition wastes, it is suggested that these materials be deposited along areas where future roadways are planned. The fact that such solid wastes are relatively inert in a landfill and do not undergo decomposition makes them excellent foundations for roadway construction. However, cities such as Muscatine, Iowa, have successfully used as much as twenty feet of well compacted mixed refuse as foundations for city streets. [1] Mixed refuse has also been used to reclaim over 10,000 acres of tidal and marshy land around New York City; this land has been used for a variety of purposes such as streets, parks, airports, industrial buildings, and large housing developments.

The construction of buildings on completed landfill often presents more problems than it is worth. It is usually only in densely populated areas where property is scarce and values high that consideration should be given to such final use of fill sites. A sanitary landfill upon which buildings are to be constructed requires more planning and more attention to operating details than other fills. Refuse should be thoroughly compacted so that the fill will settle no more than 5 to 10 percent. The final cover of two to five feet of earth should also be thoroughly compacted. Special attention should be given to the design and placement of sewer, gas, and water lines so that they will not be ruptured if settlement occurs. In addition to being flexible, underground pipes should be sealed against the possible entry of landfill gases. It is recommended that one-story buildings be placed on floating foundations, and for buildings of two or more stories, concrete or steel piling should be driven through the fill to a solid footing. Especially in instances where fill cells are penetrated by piling, allowances must be made for proper ventilation of landfill gases which will inevitably rise through these punctures and accumulate around building foundations. Basements of buildings over landfills should not penetrate the final cover material, and they should be sealed against the entry of gases. Pipes, ducts, and conduit which go through the ground floor of buildings should be sealed with permanent gastight gaskets. If buildings are to have a crawl space rather than a basement, the crawl space should be at least three feet above the finished grade and fixed louvered vent openings should be placed in the enclosing walls. It can be expected that a well seasoned landfill will support permanent uniform loading of about 2,500 pounds per square foot if successfully subjected to equivalent preload testing of at least one year. Heavier loading would normally require a special foundation of pilings.

Completed landfill sites have successfully supported agriculture, especially where irrigation is not required. Three feet of final cover is recommended, and, of course, it must be of a soil suitable for crop production. As with any landfill, proper drainage and post-fill maintenance are important considerations.

The reclamation of marginal land and abandoned mine areas with solid waste is another positive use of the sanitary landfill concept. The filling of gravel quarries, sand pits, and other similarly worked areas has been practiced for years, but such sites are often not well suited for sanitary landfilling, and special precautions must be taken to use them properly. Proximity to ground water and lack of suitable cover material are common problems with such sites. Using leachate control measures and importing cover or reducing the need for cover by preprocessing the refuse can make such sites acceptable for landfilling. The Boeing Company is involved in improving low-quality land in southern Washington by plowing milled solid waste and sewage sludge into the soil. Similar projects are in various stages of accomplishment at several locations around the country.

Strip mined areas are an eyesore and a waste of space in many parts of the country. Numerous schemes to reclaim such land have been developed around the country, and have involved almost every major city at one time or another. Many of these schemes have incorporated rail haul to transport the solid waste over long distances to the strip mined area. Plans vary from filling mine areas to the original land contours, to plowing into the soil relatively small amounts of solid waste so as not to impede plant growth.

Case History

DuPage County, Illinois, has nearly completed a particularly outstanding example of sanitary landfilling practice to meet needs of the people beyond just refuse disposal. DuPage County is immediately west of Cook County

(Chicago) and, as in other large urban areas, the pressure for recreational facilities is great and continues to grow faster than the population. The original goal of the project was to construct a recreational area, including a lake, in a worked-out gravel pit area purchased by the Forest Preserve District for that purpose. Plans called for removal of an additional 1.2 million cubic yards of gravel and clay from the old pit to make a 65-acre ground water lake. In order to maintain the water level in the lake during dry years, a separate 15-acre pond was designed to provide water by pumping to the larger lake. The amount of material to be excavated for the pond was 500,000 cubic yards. Of the total 1.7 million yards to be excavated, approximately half was commercial quality gravel valued at 75 cents per cubic yard loaded on trucks, while the remainder was clay.

The District proceeded to remove and sell the gravel, using the profits from the operation to develop the area as a park. There was a problem, however, with what to do with the clay. In 1965, the DuPage County Department of Public Works approached the District with the need for new refuse disposal sites, and a plan evolved for solving both solid waste and lake excavation problems simultaneously. The plan involved building near the lake a winter sports hill which would utilize both solid waste and the clay excavated from the lake. The hill was designed to be not only aesthetically pleasing in shape, but also suitable for winter sports. The hill is 150 feet high, and the east facing slope will provide five refrigerated toboggan runs over 800 feet long and four ski runs of various lengths. The hill contains approximately one million cubic yards each of solid waste and clay mixed with a minor amount of gravel and other soils.

The site for the hill was leveled and the base of the hill divided into eight cells of approximately five acres each. Twelve-feet wide trenches were dug completely around the cells and of a depth sufficient to extend three feet into the 55-feet thick layer of impermeable clay underlying the site. The clay layer was under a shallow covering of gravel. The trenches were filled and compacted with clay from the lake excavation to make a watertight seal between the solid waste and ground waters. Clay berms five to ten feet high by fourteen feet wide were then built on top of the clay filled trenches to prepare the cells for solid waste. The refuse was compacted in 3-foot lifts and covered with six inches of dirt-gravel or clay. Upon the completion of each lift, new clay berms were built around each cell until the hill was close to 120 feet high. Beyond this point, residential refuse was not accepted, and the hill was filled with demolition and construction debris and similar materials. Of course, as the elevation of each successive layer increased, the total area decreased, and thus the clay berms surrounding the site were moved inward to meet the desired final hill contours. The final covering of the hill assures a cap at least ten feet thick over the entire hill. The cap is composed of clay and soil. The hill was virtually completed in 1972, with only the top dressing, planting, and other final matters remaining.

The basic cost for the lake excavation and hill construction project was borne by revenues from gravel sold at 75 cents per cubic yard, and the dumping fee for refuse of 50 cents per cubic yard. The daily average amount of refuse taken at the site was about 2,000 cubic yards. Engineering, supervision, salaries, equipment, fuel, repairs, and all other expenses associated with the operation were charged to the appropriate landfill or gravel account. The equipment was purchased on a lease-purchase arrangement, with 94 to 96 percent of the rental applying toward purchase. Major equipment requirements were two crawler tractors, a front-end loader, and three scrapers for the hill construction; and two rubber-tired loaders, a scraper, and a dragline for the lake excavation.

The finished recreational area, named the Roy C. Blackwell Recreational Preserve, will be 1,400 acres in extent. It will include facilities for boating, camping, picnicking, hiking, fishing, swimming, and horseback riding in the summer months, and skiing, sledding, and tobogganing in the winter. Major earth and materials moving has been completed, with buildings, roads, final plantings, etc., being the major sub-projects yet to be completed. The area is an outstanding example of a sanitary landfill concept being put to positive use to provide an otherwise prohibitively expensive recreational facility. The project has been so successful that the district is making arrangements for a second and larger hill. Present indications are that the entire construction of this hill will be contracted out, leaving the district with only the overall administration of the project—not the operational concerns. The hill will probably be constructed with milled refuse to reduce materials handling problems, to reduce traffic on the hill during construction, and generally to improve the environmental acceptability of the operation.

The Future of Sanitary Landfilling

It would appear that sanitary landfilling will continue to experience considerable growth during the next ten to twenty years. The main reasons for this expected growth are: (1) sanitary landfilling is the only acceptable method of solid waste *disposal;* (2) sanitary landfilling offers a means of disposal which presents substantially less threat to the environment than open dumping or burning; and (3) sanitary landfilling usually has a significant cost advantage over incineration or other solid waste reduction processes. Only in densely populated areas with great haul distances to available sites does incineration become more economically attractive.

The art and science of sanitary landfilling can be expected to change during the next decade to emphasize quality operations and site design more than in the past. The basic method of landfilling, and the equipment employed to accomplish the operation, will change little, but special consideration for gas and leachate control and more emphasis on site design will be evident. Hydrogeological requirements for prospective sites will continue to become stricter, and special measures for making usable otherwise unacceptable sites will become more commonplace. The use of plastic, clay or bituminous liners to collect leachate and contain gases will become more common. Along with leachate collection, leachate treatment will be employed. A major likely change is that the sanitary landfill will be designed as a solid waste treatment system, that it will be recognized that solid wastes in a landfill can degrade, and that design can help keep the rate and extent of decomposition at acceptable and predictable levels.

It can be expected that milling and baling will play an increasingly important role in landfilling. Milling will undoubtedly become more common as recycling of solid waste becomes more widespread, for it is an especially useful way of preparing refuse for separation for salvage as well as for high-quality sanitary landfill operations. Baling will be of special usefulness as long-distance hauling to remote sites becomes more widespread.

In the future extensive recycling of paper, glass, metal, and other portions of solid wastes could remove a substantial portion of the refuse which presently finds its way into landfills. The objective for refuse disposal will include facilities for processing and reclaiming components of value for recycling, utilizing the combustible fraction in energy recovery systems and burying the residue in properly engineered, constructed and operated landfills.

Notes

1.   "From Landfill to Streets," *The American City*, p. 24, April 1966.

Appendixes

Appendix A
An Accounting System for
Sanitary Landfill Operations

Eric R. Zausner

U.S. DEPARTMENT OF HEALTH, EDUCATION, AND WELFARE
Public Health Service

ENVIRONMENTAL HEALTH SERVICE

Bureau of Solid Waste Management

The increasing costs and complexities of solid waste handling require new, more sophisticated management techniques. Data on performance and the costs of operation and ownership are essential for the use of these management tools. Hence, an adequate information system is a prerequisite to effective management. Although cost accounting represents only one part of the total information system, its design, installation, and utilization can represent the most significant step in the development of effective solid waste management.

Present information on landfill activities and associated costs is both inadequate and nonstandardized. The proposed system provides a guide to the type and quantity of information to be collected, its classification, and the method of collection. It is intended to be of use to municipal or private personnel involved in landfill operation and ownership.

Installation of a cost accounting system can aid the landfill manager in controlling the costs and performance of the operation, and also in planning for the future.

81

System Benefits

Implementation of a system such as the one described herein has several important advantages, as follows:

1. It facilitates the orderly and efficient collection and transmission of all relevant data. In fact, much of the recommended data is probably already being collected, although haphazardly and inefficiently. Hence, the added cost of implementing the system is minimal.

2. Reports are clear and concise, presenting only data required for effective control and analysis. Because they can be completed and understood by landfill personnel, operation of the system can be made almost foolproof.

3. The data is grouped in standard accounting classifications. This simplifies interpretation of results and comparison with data from previous years or other operations. In turn, this allows analysis of relative performance and operational changes.

4. The system accounts for **all** relevant costs of operations.

5. Accumulated data from the system can over a period of time lead to standards of performance and efficiency. These standards are used to control costs. They indicate what costs are high and what is causing them. The landfill supervisor may then take corrective action.

6. The system includes automatic provisions for accountability. Cost control becomes more effective when the individual responsible for cost increases can be pinpointed.

7. The collected data aids in short-range and long-range forecasting of operating and capital budgets. This facilitates estimation of future requirements for equipment, manpower, land, cash, etc., which, in turn, aids planning at all levels of management. The data is also available for later evaluation and analysis, using operations research techniques.

8. With only minor modifications, the system is flexible enough to meet the varying requirements of landfills of different size and scope.

Reports and Information Flow

The cost system is designed for medium-size sanitary land-fills. It assumes that the community or private firm has an accounting section or department to aid in preparation of the summary reports. The system also assumes that a scale is on-site. Actual measurement of solid waste quantities is possible only with scaled weights. Due to the system's flexibility, however, neither of these assumptions is critical. Only minor modifications are required to adopt the system to significantly larger or smaller operations. Due to the diversity of disposal operations no attempt will be made to suggest all of the possible variations.

The flow of information through the cost system is by means of reports (Diagram I). The eight reports transmit in-

DIAGRAM I

INFORMATION FLOW DIAGRAM

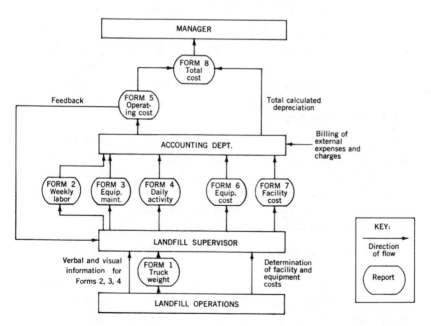

formation from the field where data is recorded to the point of data use — various levels of supervision and management.

The reports are most easily classified as those that are prepared daily and those that are prepared at less frequent intervals. Cost control, responsibility accounting, and the preparation of periodic summary statements cannot be accomplished without the daily recording of all pertinent activity and cost information. Data not recorded daily is not retrievable at some later date.

Daily truck weight record (Form 1). This form records the quantities, sources, and types of solid wastes delivered to the site. Cover material, if it is delivered from off-site, may be recorded on this form. If the cover material is acquired on the site, the "cover material" column may be deleted or a daily estimate of cubic yards may be recorded. The information is recorded manually on this form for the entire day. If the landfill has a scale that automatically records the weight data, that part of the form would be replaced by the weight ticket. Each delivery is recorded separately by the weighmaster. A second weighing of the empty truck may be taken, or the vehicle's tare weight (as determined by a licensing agency, etc.) may be substituted.

Truck identification, load weight, and solid waste type are useful in billing private concerns and others for the privilege of using the landfill. The record of truck delivery times and frequency, along with truck identification, load weight, and waste type, are an important aid to operation and control of a municipal collection system.

Weekly labor ticket (Form 2). This record of labor activity is recorded in duplicate at the landfill. One copy is forwarded to the payroll department for determining weekly wages. The other copy is used by the landfill supervisor for computing total labor hours and the assignment of these hours to the landfill's various activities.

Monthly equipment maintenance record (Form 3). This form facilitates the detailed collection of equipment operation and associated cost data. A separate sheet is used for each piece of equipment at the site. Daily entries are made as appropriate. The form is used for an entire month.

FORM 1

DAILY TRUCK WEIGHT RECORD

SITE: _____   DATE: _____ / _____ / _____

SIGNATURE: _____

Truck ident.	Time*	Wastes		Weight in	Weight out (tare weight)	Amount delivered	
		Source†	Type‡			Solid waste	Cover mat'l
Totals	X	X	X	X	X		

Instructions: To be completed by weighmaster for each truck, each time it makes a delivery. If scales are not available, an estimate of cubic yards is acceptable.

* Only record time at 15-minute intervals.

† Sources: R = residential, I = industrial, C = commercial, A = agricultural, etc.

‡ Types: H = household refuse, D = demolition/construction wastes, T = tires, B = bulky waste — furniture, refrigerators, etc.

BSWM (7/69)

FORM 2

WEEKLY LABOR TICKET

SITE: _____

SIGNATURE: _____

DATE: ___ / ___ / ___

Employee ident.	Day 1		Day 2		DAY 3		DAY 4		DAY 5		DAY 6		DAY 7		Indiv- idual totals	Note causes of absences, extra hrs. to be paid, etc.
	Job	Hrs.	Job	Hrs.	Job	Hrs.	Job	Hrs.	Job	Hrs.	Job	Hrs.	Job	Hrs.		
Totals	X		X		X		X		X		X		X			XXXXXXXXXXXXXXXXX

Instructions: Landfill supervisor to complete this form daily. List all employees separately including temporary help. "Hrs." refers to hours worked daily. "Job" refers to the job description.* At the end of each week, forward one copy to the payroll department and retain the original for further use.

* TD = tractor driver, W = weightmaster, ME = equip. maint, MB = building maint, etc. Tractor driver activity may be divided into spread and compact refuse(s) and cover operation (c).

BSWM (7/69)

FORM 3

MONTHLY EQUIPMENT MAINTENANCE REPORT

SITE: _____   PERIOD: from _____ to _____

EQUIP. IDENT.: _____

Day	Hours operated	Fuel cost	Maintenance						
			Type of repair or service	Hours down	Labor hours	Labor cost	Parts cost	External cost	Total cost
1									
2									
3									
4									
5									
6									
7									
8									
9									
10									
11									
12									
13									
14									
15									
16									
17									
18									
19									
20									
21									
22									
23									
24									
25									
26									
27									
28									
29									
30									
31									
Totals			XXXX						

Instructions: To be completed by the landfill supervisor at the end of each day. Entries under maintenance should be made only as they occur. This sheet is only for one piece of equipment.

BSWM (7/69)

Daily activity summary (Form 4). This form summarizes the truck deliveries, solid waste quantity disposed of, man-hours worked, machine hours utilized, cover material used (if measured or estimated), and miscellaneous expenses incurred. Since this provides a continuous cover material inventory, it is useful in keeping the landfill supervisor advised as to his present status and when more cover material will be required. The form is used for an entire month. It is completed at the end of each day by the landfill supervisor. At the end of the month, it is forwarded to the accounting department.

Less Frequently Prepared Reports

These reports may be prepared as often as desired. The Operating Cost Report is used for control purposes. Hence, the more frequently it is prepared (perhaps even weekly), the more useful it would be. The remaining reports are summary reports (preparation quarterly would be sufficient).

Landfill operating cost report (Form 5). This report summarizes the landfill's operations. It is compiled from all the daily tickets. As a summary of the landfill's total operating costs, it may be used to hold the supervisor responsible for any adverse trends in costs. In addition, the calculated unit cost and efficiency factors are helpful to the supervisor in analyzing these adverse cost trends and controlling them. This is more fully discussed under **system utilization.**

If most of the expenses incurred in operating the facility are billed directly to the municipality, the necessary cost data required for preparation of this form would not normally be available to the supervisor. Under these circumstances, the supervisor should forward the information he has collected (the daily tickets) to the accounting department. This department will then compile the complete Operating Cost Report and send a copy back to the landfill supervisor.

A more detailed breakdown of the expenses that should be included under each category is in Form 5a. Form 5a need not be prepared; it is only shown to indicate the relevant cost groupings.

FORM 4

DAILY ACTIVITY SUMMARY

SITE: _____   DATE: _____ / _____ / _____

SIGNATURE: _____

DAY	Solid waste		Cover material				Man hrs.	Machine hrs.		Expense*		Site hrs.
	Loads	Tons	Begin	Rec'd	Used	Remain		Use	Down	$	Type	
1												
2												
3												
4												
5												
6												
7												
8												
9												
10												
11												
12												
13												
14												
15												
16												
17												
18												
19												
20												
21												
22												
23												
24												
25												
26												
27												
28												
29												
30												
31												
Totals			X			X					X	

Instructions: To be completed by landfill supervisor at the end of each day. Some of data is to be summarized from Forms 1 and 2. Record cover material in either tons or cubic yards. Today's beginning cover material equals yesterday's remaining.

* Record only miscellaneous expenses that would not be recorded by the accounting department and are incurred at the site.

BSWM (7/69)

FORM 5

LANDFILL OPERATING COST REPORT

SITE: _____   PERIOD OF REPORT: from _____ to _____

	Data	Actual this period	% var. from budget	% var. from last period	% var. this period last year
Totals	Total tons received				
	Total operating cost				
	Total operating cost/ton				
Unit costs*	Labor/ton				
	Cover material/ton				
	Equip. operation/ton				
	Overhead/ton				
Efficiency factors*	Cover material util.				
	Overtime hours/total labor hours				
	Labor efficiency				
	Equip. % downtime				
	Equipment utilization				
	Equipment efficiency				

Instructions: To be completed by accounting department from Forms 1, 2, and 3. One copy to landfill supervisor.

* Calculations: Unit cost = aggregate cost ÷ tons solid waste received. Note that cover material unit cost is cover material cost ÷ tons of solid waste received.

Cover material utilization = cover material used ÷ tons of solid waste received.

Labor efficiency = tons received ÷ labor hours.

Equipment utilization = tons received ÷ equipment hours.

Equipment efficiency = equipment cost ÷ equipment hours.

Equipment % downtime = total hours down ÷ total equipment hours.

BSWM (7/69)

FORM 5a

OPERATING COST CLASSIFICATIONS

Labor costs

Include all wages at base pay, plus all overtime pay, and labor fringe benefits.*

Cover material costs

Include all costs for the delivery of cover material. This category may be excluded if cover material is not delivered from off the site. The cost of labor, equipment, and overhead required to obtain on-site cover may be substituted.

Machine operating costs

Include oil, gasoline, grease, equipment repairs, and maintenance.

Overhead costs

Include all utilities, supervisor's salary, building repairs and maintenance, liability and property insurance, and charges from other departments.

---

*Labor fringe benefits include group insurance, pension costs, social security contributions, vacation costs, etc., whether actually budgeted to the operating agency or absorbed in municipal general funds.

Equipment and facility cost reports (Forms 6 & 7). These two reports are compiled at the landfill site or wherever the data is available. They are then updated only when additional equipment or facilities are acquired. The periodic depreciation charges are then computed and posted by the accounting department.

Landfill total cost summary (Form 8). This report summarizes all the activities and costs incurred by the landfill during the period. It is compiled from data available in present and past Operating Cost Reports and the depreciation data available on the Facility and Equipment Cost Reports.

Report Flow Summary

A brief summary may help to put the system in perspective. Operating reports are generated daily at the landfill site and transmitted periodically to the accounting department. The accounting department combines these reports with additional information it accumulates to produce total operating costs.

System Utilization

Now that the actual system has been described and one possible set of forms illustrated, utilization must be discussed. Only with efficient and intensive utilization of the information generated, can the additional time, effort, and money required to implement and maintain the system be justified.

All the factors which affect the quality and effectiveness of sanitary landfill operations can be translated into costs. Extent of cover material use, the size of the face, degree of compaction, litter control and dust control, among others, determine how good a job of sanitary landfilling is performed, and they are more costly than simple operation of an unattended open dump. Cost control at a landfill does not call for economizing at the expense of quality. To the contrary, once a level of acceptable operation has been determined along with the attendant costs, the cost control system can help the supervisor maintain that level of operation.

The routine control of costs is slightly more complicated. Effective cost control has two prerequisites: recognition of

FORM 6

LANDFILL EQUIPMENT COST REPORT

SITE: _____

DATE: _____ / _____ / _____

(for use by acctg. dept. only)

Type	Model no.	Model year	Mfg. name	Date of purch.	Cost	Est. life	% Time used by other dept.	Annual depreciation	Monthly depreciation
Totals	X	X	X	X		X	X		

Instructions: To be filled out by accounting dept. or supervisor. "Est. life" should be based on supervisor's estimate of remaining life. Use of equipment by other dept. should be based on percent of time (working day) equipment is away from the landfill. Depreciation may be on a straight-line or accelerated basis.

13

BSWM (7/69)

FORM 7

LANDFILL FACILITY COST REPORT

SITE: _____

DATE: ___ / ___ / ___

(for use by acctg. dept. only)

Item or category		Description	Date put in use	New cost	Est. total life	Other comments	Annual depreciation	Monthly depreciation
	All land						XXXXXX	XXXXXX
Improvements	Roads							
	Lights							
	Fences							
	Surveys							
	Other							
Facilities	Scales							
	Garages							
	Buildings							
	Other							
Totals		X	X		X	X		

14

Instructions: To be completed by supervisor or accounting dept., if they have data available. "Est. total life" should be based on remaining life as estimated by the supervisor. Land purchased subsequent to the original land purchase should be included. Depreciation may be either straight-line or on an accelerated basis.

BSWM (7/69)

FORM 8

LANDFILL TOTAL COST REPORT

SITE: _____

PERIOD OF REPORT: from _____ to _____

Data	For this period	Budget—this period	Year to date	Budget—year to date
Tons of solid waste received				
Total operating cost				
Total depreciation cost				
Total cost				
Operating cost per ton				
Depreciation cost per ton				
Total cost per ton				

15

Instructions: To be completed by the accounting department, when requested or periodically, from data available in operating cost report or capital cost reports. Copies sent to the city manager (or his equivalent).

BSWM (7/69)

excessive costs and identification of responsibility for the increased costs. By comparing present unit costs with the currently budgeted unit costs and the actual unit costs of the previous period and the same period last year, some determination can be made of whether present costs are excessive. The determination of responsibility is facilitated by the efficiency factors. The system described allows both of these critical factors to be determined. Corrective action may then be effectively initiated.

At the highest level of management, the Total Cost Report indicates whether costs are excessive, in which case the supervisor of the particular sanitary landfill can be held responsible. The supervisor, in turn, can use the cost system to determine the cause of increased costs. He may trace the increased costs to the particular cost element, and possibly to the employee, piece of equipment, or method of operation responsible. All of the needed data is in Form 5 (the Operating Cost Report). To aid the supervisor in the analysis of Form 5, a decision tree may be used (Diagram II). It illustrates the methodology required to analyze the cause of increased costs. For clarity, a hypothetical situation will be examined.

Let us assume that the landfill supervisor receives his copy of the Operating Cost Report from the accounting department. His analysis of the data starts at the extreme left of Diagram II. Quite obviously, the first question to be answered is whether any analysis is required. If total operating cost per ton is less than or equal to the budget, the answer is No. However, if total cost per ton is greater than the budget, additional analysis is indicated. Next, it is desirable to isolate the cost element which is abnormally high. It may be one or more of the four shown (labor, cover material, equipment, or overhead). Let us assume that only labor cost per ton is higher than its budgeted amount. (We are now on the uppermost branch.) We must determine why labor cost per ton increased, so that corrective action can be taken. Several factors are listed which may be relevant. Assume "overtime hours per total labor hours" is excessive. This implies that either scheduling is poor, there is a temporary peak load, the

DIAGRAM II

DECISION TREE FOR SANITARY LANDFILL
COST VARIANCE ANALYSIS

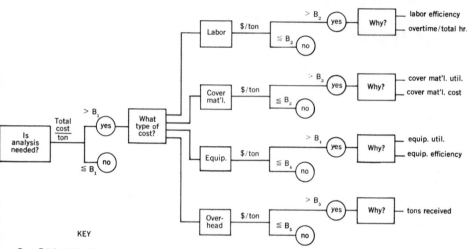

KEY

B_1 = Total cost/ton budgeted

B_2 = Labor cost/ton budgeted

B_3 = Cover mat'l. cost/ton budgeted

B_4 = Equip. cost/ton budgeted

B_5 = Overhead cost/ton budgeted

employees are working overtime when not required, or perhaps more regular employees are required. Under any of
these circumstances, corrective action could be initiated by
the supervisor.

This example is an over-simplification of actual operations.
Nevertheless, it does illustrate the use of the decision tree
(Diagram II) and more importantly, the methodology needed

to pinpoint and correct factors which might have caused increased costs.

In addition to routine cost control, the data collected is useful in doing special analyses of trends in composition and quantity of wastes, peak load hours, on-site waiting times, and equipment evaluations These and other quantitative evaluations can improve performance and reduce the costs of operation.

U.S. GOVERNMENT PRINTING OFFICE : 1970 O—380-584

Appendix B
Model Sanitary Landfill
Operation Agreement

This agreement made and entered into this _____ day of 19__,
by and between _____ (a City or a
Village, County, etc.) organized under the laws of the State of _____,
hereinafter referred to as the City (Village, etc.) and _____
_____ (a Corporation or a Partnership,
Proprietorship, etc.) organized under the laws of the State of _____
and having its principal place of business at _____
_____ hereinafter referred to as
the Contractor.

WITNESSETH:

Whereas, the Contractor is qualified to operate a sanitary landfill for the
disposal of solid waste in accordance with the Instructions to Bidders; and

Whereas, the City desires the Contractor to operate the area designated
to be used for a sanitary landfill operation;

Now therefore, in consideration of the mutual covenants and agreements
contained herein, and of the consideration paid by the City to the Contractor,
as hereinafter set forth, the City and the Contractor hereby agree as follows:

1. *Disposal Site.* All solid waste shall be disposed of at the location(s)
specified herein, same being the property under the control of the City (or
Contractor), and more specifically described as follows:

[Insert Legal Description]

(It should be recognized that the site may be provided either by the
local government authority or the contractor).

2. *Materials to be Disposed of.* The Contractor shall accept all solid waste
created within the jurisdiction of the City or that for which the City has
accepted responsibility upon payment of fees as scheduled. Toxic, volatile,
and other hazardous materials must be clearly identified to allow for special
precautions to be taken during the disposal operation.

3. *Inspection.* The City may make inspections of the sanitary landfill
site through designated personnel during business hours.

Source: *Sanitary Landfill Operation Agreement and Recommended Standards for Sanitary
Landfill Design and Construction,* National Solid Wastes Management Association and U.S.
Department of Health, Education, and Welfare, Bureau of Solid Waste Management, Report
no. SW–20ts, (1969).

4. *Operation of Site.* The Contractor shall have the exclusive right and responsibility for the operation of the disposal site in accordance with the provisions of this Agreement for the term of this Agreement and any extension thereof.

5. *Compliance with Laws.* The Contractor shall operate the disposal site in compliance with all applicable laws, ordinances and regulations; including the Solid Waste Disposal Act of the State of _____ , the rules and regulations of the State Board of Health and the City and/or County Board of Health, and the Ordinances of the City and/or County, copies of each being attached hereto and are hereby made a part of this Agreement. Copies of all such laws, ordinances and regulations shall be furnished to the Contractor by the City and shall include new legislation and amendments to existing legislation.

In the event that there are no statutes and/or ordinances regulating the disposal of wastes, then the Contractor shall design and operate said disposal site in full compliance with the *Recommended Standards for Sanitary Landfill Operation,* or revisions thereof (Exhibit C attached). Said *Standards for Sanitary Landfill Operation* shall be revised by the City only, but the Contractor is not hereby precluded from submitting written recommendations thereon to the City.

6. *Labor and Equipment.* The Contractor shall furnish all labor, tools and equipment necessary for the operation of the site and shall be responsible for all required maintenance thereof. Supervision by an experienced and qualified person shall be provided at all times the sanitary landfill is open for use or operation.

7. *Service Facilities.* The Contractor shall construct and maintain, at his expense, any facilities, improvements and buildings within the site necessary for the operation of the site.

([To be included if the site is City property:] The use of such land within the site shall be made available to the Contractor free of charge for the period of this Agreement or any extension thereof. At the expiration of this Agreement all permanent structures and improvements shall become the property of the City or shall be removed by the Contractor at the option of the City. [If permanent structures and improvements become the property of the City, there should be some provision for compensation to the Contractor, such as book value or fair market value. If the Contractor is required to remove such structures and improvements then he should be paid to do so.])

8. *Offsite Improvements.* The City at its expense agrees to provide all required offsite improvements including any improvements required on public streets or roads, drainage facilities, etc.; and shall provide to the site all required utilities including adequate power and water supplies. (If any of this work is

to be performed by the Contractor, it should be included in a separate contract with detailed engineering plans.)

9. *Charges for Utilities.* The Contractor agrees to pay the normal and standard charges for all water, electrical power, natural gas, and phone service utilized at the site. (If any of the services are to be provided free of charge by the City, this section should be modified accordingly.)

10. *Salvage.* Neither scavenging nor salvage operations shall be permitted at the operating face of the sanitary landfill. (If salvage is permitted, then insert: Salvage operations shall be conducted at a location separate from the operating face of the landfill by persons licensed by the City so as not to interfere with the operation of the sanitary landfill by the Contractor.)

11. *Title to Waste.* Title to waste shall vest in the owner of the fee simple estate as it is deposited.

12. *Completion of the Site.* Upon completion of disposal operations, the site shall be contoured and finished in accordance with the Approved Final Plan, which is attached hereto and is hereby made a part of this Agreement and the Operating Standards. Any changes of the Approved Final Plan must be agreed to by both the City and the Contractor. The liability of the Contractor under this agreement shall cease upon acceptance of the site by the City.

13. *Compensation.* Compensation shall be paid pursuant to the attached schedule. (A model schedule could be attached here and provide for payment by weight or volume. A minimum charge should be set out. Hazardous materials would be handled on a mandatory basis on fees paid for each type and quantity handled. Experience should soon establish fees for such materials.)

(If materials are to be accepted from users other than those paid for by the City, a similar schedule of prices which the Contractor can charge these users should be established. There should be clear provisions regarding the distribution of such fees to the Contractor and/or the City.)

14. *Change in Regulations.* In the event that compliance with subsequent statutes, ordinances and/or rules and regulations results in change in operating costs, the parties hereto agree to renegotiate this Agreement so that the compensation herein shall reflect such change.

15. *Change in Sanitary Landfill Site.* In the event that the parties hereto mutually agree to transfer said sanitary landfill operations to another site, or additional sites, then this Agreement shall be renegotiated to reflect any changes required by said transfer or addition, including, but not limited to increases in compensation due to increased operating costs.

16. *Change in Cost of Doing Business.* The fees and/or compensation payable to the Contractor for the second and subsequent years of the term hereof shall be adjusted upward or downward to reflect changes in the cost

of doing business, as measured by fluctuations in the consumer price index published by the U.S. Department of Labor, Bureau of Labor Statistics, for the _____ area. At the start of the second year and every six (6) months thereafter, the fees and/or compensation to the Contractor shall be increased or decreased in a percentage amount equal to the net percentage change in the said consumer price index computed as follows:

Beginning with the first month of the second year the net change shall be the difference between the said consumer price index during the first month of this Agreement and the last month of the first year.

Beginning with the seventh month of the second year of this Agreement and every six months thereafter, the net change in the said consumer price index shall be the change for the last preceding six month period.

17. *Term.* The initial term of this Agreement shall be for the _____ -year period beginning _____ , 19 _____ , and ending _____ , 19 _____ . The initial _____ year term of this Agreement shall be extended for successive additional _____ -year terms, unless either party notifies the other party in writing as hereinafter provided, not less than ninety (90) days prior to the expiration of the initial _____ year term or of any _____ year extension thereof of its intention to terminate this Agreement.

18. *Performance Bond.* The Contractor shall furnish a Performance Bond for the faithful performance of this Agreement, said bond to be executed by a surety company licensed to do business in this State, and to be in a penal sum equal to one-half the minimum fee to be paid to the Contractor by the City for the first year of this Agreement, and for each year thereafter to be in the penal sum of _____ percent of the total compensation paid by the City to Contractor for the last preceding year of this Agreement, but in no case to exceed 100 percent of the compensation paid during the prior year of this Agreement. Said Performance Bond shall be furnished annually by the Contractor on the first day of each year of this Agreement or any extension thereof, and shall indemnify the City against any loss resulting from any failure of performance by the Contractor, not exceeding, however, the penal sum of the bond.

19. *Payment Bond.* The Contractor shall, within ten (10) days of the execution of this agreement, deliver or cause to be delivered to the City a bond in the amount of $ _____ executed by a surety company licensed to do business in this State, guaranteeing payment of wages to all employees of the Contractor at the site or sites and the cost of all supplies, materials, and insurance premiums required in fulfilling this Agreement.

20. *Indemnify.* The Contractor hereby binds himself to indemnify and hold harmless the City from all claims, demands and/or actions, legal and/or equitable, arising from the Contractor's operation of all disposal sites herein above described.

Proof of such insurance shall be furnished by the Contractor to the City by certificates of insurance, with a minimum cancellation time of ten (10) days, said time to commence after delivery of said notice to the City at the address shown above.

(Liability insurance policies approved by the City as to type and coverage, may be required by the City as a part of the indemnity provisions of this Agreement. If policies such as automobile liability, general liability or owner's protective liability are required, the type and amount of coverage should be clearly spelled out in this section. Proof of such insurance shall be furnished by the Contractor to the City by certificates for each policy, with each policy having a minimum notice of cancellation to the City of ten (10) days.)

21. *Workmen's Compensation.* The Contractor shall carry in a company authorized to transact business in the State of _____ , a policy of insurance fulfilling all requirements of the Workmen's Compensation Act of said State, including all legal requirements for occupational diseases. (Would not apply in monopoly States.)

22. *Standard of Performance.* In the event that the Contractor shall fail to dispose of materials herein provided to be disposed of for a period in excess of five (5) consecutive days, and provided such failure is not due to war, insurrrection, riot, Act of God, or any other cause or causes beyond the Contractor's control, the City may, at its option, after written notice to the Contractor as provided hereinafter, take over and operate any or all of the Contractor's equipment used in the performance of this Agreement, and provide for such operation until such matter is resolved and the Contractor is again able to carry out its operations under this Agreement. Any and all operating expense incurred by the City in so doing may be deducted by it from compensation to the Contractor hereunder.

During such period, the liability of the City to the Contractor for loss or damage to such equipment so used shall be that of a bailee for hire, ordinary wear and tear being specifically excepted from such liability. The liability of the Contractor to third persons shall cease and all claims or demands arising out of the operation and/or control of the site or sites shall be directed solely to the City.

Provided, however, if the Contractor is unable for any cause to resume performance at the end of thirty (30) calendar days, all liability of the City

under this contract to the Contractor shall cease and the City shall be free
to negotiate with other contractors for the operation of said site or sites. Such
operation with another contractor shall not release the Contractor herein of
its liability to the City for such breach of this agreement. In the event that
this contract is negotiated with another contractor(s), third-party liability of
the Contractor herein shall terminate insofar as same arises from tortious
conduct in operation and control of sites.

23. *Arbitration.* Any controversy or claim arising out of or relating to
this Agreement, or breach thereof, shall be settled by arbitration in accordance
with the Rules of The American Arbitration Association, and judgment upon
the award rendered by the arbitrators may be entered in any court having
jurisdiction thereof. Such controversy or claim shall be submitted to one arbi-
trator selected from the National Panel of The American Arbitration Association.

(Arbitration Statutes are in effect in the following States: Arizona,
California, Connecticut, Florida, Hawaii, Illinois, Louisiana, Maryland, Massachusetts, Michigan, Minnesota, New Hampshire, New Jersey, New York, Ohio,
Oregon, Pennsylvania, Rhode Island, Texas, Washington, Wisconsin, and Wy-
oming. Common law governs in all other States.)

24. *Landfill Inspection.* To insure that the detailed specifications, regulations, and laws for the operation of a sanitary landfill are complied with, the
landfill site and operation shall be inspected a minimum of once a month
during the term of this Agreement by a representative of the City to assure
compliance.

25. *Assignment.* No assignment of this Agreement or any right occurring
under this Agreement shall be made in whole or part by the Contractor without the express written consent of the City, and in event of any assignment
the assignee shall assume the liability of the Contractor.

26. *Books and Records.* The Contractor shall keep daily records of
wastes received and the City shall have the right to inspect the same insofar
as they pertain to the operation of the sanitary landfill site(s) for the purpose
of analysis of the financial condition of said operation. The records should
include the type, weight and volume of solid waste received, the portion of
the landfill used (determined by cross section and survey) and any deviations
made from the plan of operation and equipment maintenance and cost records.
The contractor shall submit a proposed record and accounting system for
approval. All information so obtained shall be confidential and shall not be
released by the City unless expressly authorized in writing by the Contractor.
A recommended set of acceptable records is included in Exhibit D.

27. *Bankruptcy.* This Contract shall terminate in the case of bankruptcy, voluntary or involuntary, or insolvency of the Contractor. The time of termination in the event of bankruptcy shall be the day and time of the filing of the petition in Bankruptcy.

28. *Number of Copies.* This Agreement may be executed in any number of counterparts, all of which shall have the full force and effect of an original for all purposes.

29. *Law to Govern.* This Agreement shall be governed by the laws of the State of _____ , both as to interpretation and performance.

30. *Modification.* This agreement constitutes the entire agreement and understanding between the parties hereto, and it shall not be considered modified, altered, changed or amended in any respect unless in writing and signed by the parties hereto.

31. *Right to Require Performance.* The failure of the City at any time to require performance by the Contractor of any provisions hereof, shall in no way effect the right of the City thereafter to enforce same. Nor shall waiver by the City of any breach of any provisions hereof be taken or held to be a waiver of any succeeding breach of such provision or as a waiver of any provision itself.

32. *Illegal Provisions.* If any provision of this Agreement shall be declared illegal, void, or unenforceable, the other provisions shall not be effected, but shall remain in full force and effect.

33. *Notice.* A letter addressed and sent by certified United States mail to either party at its business address shown hereinafter shall be sufficient notice whenever required for any purpose in this Agreement.

34. *Effective Date.* This Contract shall become effective and the City shall begin delivery of the solid waste to the Contractor _____ days after the date of execution hereof.

City: _____

_____

_____

_____

Contractor: _____

_____

_____

_____

IN WITNESS WHEREOF, the City and the Contractor have executed this Agreement as of the day and year first above written.

Approved to Form _____

City Attorney

City of   _____

A municipal corporation of the State of

_____

By _____

By _____

By _____

(Name of Contractor)

By _____

By _____

(Sealed, witnessed, and/or notarized as required by the laws of applicable State.)

FEE SCHEDULE

(Alternate methods of charge)

1. $_____ per ton of solid waste

2. $_____ per yard of compacted solid waste

3. $_____ per yard of uncompacted solid waste

4. $_____ minimum fee per load

5. $_____ per ton of solid waste consisting solely of material
         such as bricks, concrete, dirt, etc.

6.   The city shall pay to the contractor a minimum fee of $ _____
     for each year, or extension, of this agreement.

     Toxic, volatile or other hazardous materials requiring special handling
shall be clearly marked by the City and, upon payment of mutually agreed
upon fees, shall be disposed of by the Contractor pursuant to the terms of
Item 2 of the Contract.

     The Contractor shall submit billings to the City at the close of business
at the end of each month for all other waste placed in the sanitary landfill and
the City shall pay the Contractor on or before the tenth day of the following
month, payments to be mailed to the Contractor at the address shown above.

Bibliography

1. Andersen, J. and J. Dornbush. "Influence of sanitary landfill on ground water quality." *Journal of American Water Works Association,* 59 (4): 457–470, April 1967.
2. Anderson, R.J. "The public health aspects of solid waste disposal." *Public Health Reports,* 79 (2): 93–96, February 1964.
3. "Back to burning of refuse?" *Municipal and Public Services Journal,* 78: 1876, June 10, 1966.
4. Barnes, S. "Disposal gap." *Machine Design,* 38: 144–150, March, 1961.
5. Bendixen, T.W. *Review of the national solid wastes program.* Cincinnati: U.S. Department of Health, Education, and Welfare, 1967.
6. Bishop, W., R. Carter, and H. Ludwig. "Gas movement in landfilled rubbish." *Public Works:* 96 (11): p. 64, November 1965.
7. Bjornson, B.F. and M.D. Bogue. "Keeping sanitary landfill sanitary." *Public Works,* 92 (9): 112–114, September 1961.
8. Black, R.J. "A review of sanitary landfilling practices in the United States." In *Proceedings,* Third International Congress, International Research Group of Refuse Disposal, May 24–29, 1965.
9. Booth, E. and E. Carlson. "Rubber tires work well on sanitary landfills; use in Bismark, North Dakota." *The American City,* 81: 98–99, July 1966.
10. Boskoff, A. "Depreciation claim allowance on disposal site airspace." *Solid Waste Management Refuse Removal Journal,* 14 (5): 47, 159, May 1971.
11. "California task force completes its report." *Solid Waste Management /RRJ,* 13 (3): 10, March 1970.
12. Cannella, A.A. "The refuse disposal problem." *Public Works,* 99 (2): 116, February 1968.
13. Caterpillar Tractor Company. "Questions and answers about sanitary landfills," 1967.
14. "Community tries to block reclamation project." *Refuse Removal Journal,* 9 (9): 34–35, September 1966.
15. "Controversy still rages over Japanese 'building blocks.'" *Solid Waste Management/RRJ,* 14 (3): 10, March 1971.
16. Coppa, R.B. "How to start a sanitary landfill." *The American City,* 83 (3): 85, March 1968.
17. "County battles for landfill sites." *Solid Waste Management/RRJ,* 12 (3): 26, March 1969.
18. County of Los Angeles, Dept. of County Engineer. *Development of construction and use criteria for sanitary landfills.* Cincinnati: U.S. Dept. of Health, Education, and Welfare, 1969.

19. "Court decision may affect landfill." *Refuse Removal Journal,* 10 (5): 23, May 1967.

20. Cowman, H. "Town turns landfill over to private contractor." *Solid Waste Management/RRJ,* 11 (8): 14, August 1969.

21. "Create joint landfill to serve nine contractors." *Refuse Removal Journal,* 9 (9): 24, September 1966.

22. Crites, R.R. "Model town planning develops model landfill." *Solid Waste Management/RRJ,* 13 (8): 14, August 1970.

23. Culham, W.B. "Equipment needed for a sanitary landfill." *The American City,* 84 (1): 100, January 1969.

24. Cummins, R.L. *Effects of land disposal of solid wastes on water quality.* Cincinnati: U.S. Department of Health, Education, and Welfare, 1968.

25. Davis, K. "Planned landfills cut costs and complaints." *The American City,* 83 (12): 102, December 1968.

26. "Dealing with domestic refuse." *Engineering,* 203 (5267): 499–508, March 1967.

27. Department of Sanitary Engineering. *Kenilworth model sanitary landfill; interim report on a solid waste demonstration project, December 1967–January 1969.* Washington, D.C.: U.S. Government Printing Office, 1969.

28. "Detroit looks to transfer stations and landfill for tomorrow's waste management." *Solid Waste Management/RRJ,* 14 (4): 28, April 1971.

29. "Disposal near water banned by Ohio law," *Solid Waste Management/RRJ,* 11 (7): 42, July 1968.

30. Drott Manufacturing Corporation. *Sanitary fill–cost and operational data on the Drott Bullclam shovel method of sanitary landfill.* Wausau, Wis., 1967.

31. Elliott, R.S. "Texans choose sanitary landfill over burning dump." *Public Works,* 97: 92, July 1966.

32. Evans, H., Jr. "A new idea in landfill operation." *The American City,* 82 (3): 114, March 1967.

33. "First American test of pulverizing ends after three years." *Solid Waste Management/RRJ,* 13 (7): 10, July 1970.

34. "First trench-type fill opened in Oregon county." *Solid Waste Management/RRJ,* 14 (1): 89, January 1971.

35. Fleming, R.R. "Solid waste disposal; sanitary landfill." *The American City,* 81: 101–104, January 1966.

36. Frye, J. *Evaluating sanitary landfill sites in Illinois.* Environmental Geology Notes, no. 27. Urbana, Ill.: Illinois State Geological Survey, August 1969.

37. "Fundamentals of sanitary landfill operation." *Public Works,* 95 (12): 88, December 1964.

38. "Garbage: rosy new future as raw material." *Chemical Engineering,* 75 (9): 82–84, April 22, 1968.

39. "Geophysical instruments help locate landfill sites." *The American City,* 83 (5): 48, May 1968.

40. Gershowitz, H. "First national landfill group to be formed." *Solid Waste Management/RRJ,* 14 (5): 63, May 1971.

41. ———. "Model landfill contract recommends very stringent performance bond." *Solid Waste Management/RRJ*, 11 (7): 16, July 1968.

42. "Giant railhaul landfill plan." *Solid Waste Management/RRJ*, 11 (3), March 1968.

43. Gilbertson, W.E. "Scope of solid waste program." *American Society of Civil Engineers, Journal of Sanitary Engineering Division*, October 1966.

44. Golueke, C. *Comprehensive studies of solid waste management: Abstracts and excerpts from the literature.* Sanitary Engineering Research Laboratory, no. 69–7. Berkeley: University of California Press, July 1969.

45. Golueke, C. and P. McGauhey. *Comprehensive studies of solid waste management, Second annual report.* Sanitary Engineering Research Laboratory, no. 69–1. Berkeley: University of California Press, January 1969.

46. Greenspan, S.G. "Accelerated depreciation offers greater savings." *Solid Waste Management/RRJ*, 14 (2): 20, February 1971.

47. Hamlin, G.H. "Propose train haul to desert landfill." *Refuse Removal Journal*, 10 (3): 10, March 1967.

48. Hanke, J.J. and H.D. Kube. "Industry action to combat pollution." *Harvard Business Review*, 44: 49–62, September 1966.

49. Harza Engineering Company. *Land reclamation project; an interim report.* Cincinnati: U.S. Dept. of Health, Education, and Welfare, 1968.

50. Hattery, G.S. "Get out of the dumps." *The American City*, 82 (6): 100, June 1967.

51. Heaney, F.L. and C.V. Keane. *Solid waste disposal.* Boston: Camp, Dresser, and McKee, Consulting Engineers, undated. 6 p.

52. Hernandez, G. "Deep hole method extends landfill use." *The American City*, 82 (3): 17, March 1967.

53. Holheuzer, O. "Steel wheel dozer improves landfill compaction." *Public Works*, 98 (4): 115, April 1967.

54. Hopson, R.S. "From landfill to heliport." *The American City*, 83 (10): 42, October 1968.

55. Horbitz, W.E. "Looking to the future with a regional refuse disposal plan." *Public Works*, 98 (6): 120–21, June 1967.

56. "How to win friends for landfill." *Refuse Removal Journal*, 10 (4): 10, April 1967.

57. Hughes, G.M., R.A. London, and R.N. Farvolden. *Hydrogeologic data from four landfills in Northeastern Illinois.* Environmental Geology Notes, no. 26, Urbana, Illinois: Illinois State Geological Survey, March 1969.

58. Jenkins, H.W. "The monster." *The American City*, 85 (9): 146, September 1970.

59. Kandle, R. "Disposal of 21,100 tons daily, New Jersey dilemma." *Refuse Removal Journal*, 9 (10): 26–27, October 1966.

60. Kennedy, J.C. "Current concepts in the disposal of solid wastes." *Journal of Environmental Health*, 31 (2): 149–153, September–October 1968.

61. Klee, A.J. "Let Dare make your solid-waste decisions." *The American City*, 85 (2): 100, February 1970.

62. Klein, S. "New building constructed on sanitary landfill." *Public Works*, 99 (10): 125, October 1968.

63. Kletter, H. "Planning ahead helps win approval for disposal site." *Solid Waste Management/RRJ,* 12 (7): 26, July 1969.

64. Koch, A.S. "Plan for 8,150 tons daily at landfill." *Refuse Removal Journal,* 10 (8): 14, August 1967.

65. "Landfill conversion attracts new industries to Ohio town." *Refuse Removal Journal,* 10 (9): 82, September 1967.

66. *Landfill survey.* New York University, Sanitary Research Laboratory, 1967.

67. Larkin, J. "The economic advantages of beautified and properly engineered landfill." *Solid Waste Management/RRJ,* 14 (5): 46, 78, 82, 156, May 1971.

68. Mackay, B.B. "Louisiana Court says operations sites not a nuisance." *Solid Waste Management/RRJ,* 14 (1): 36, January 1971.

69. Mailey, H. "Landfill: from eyesore to asset." *Public Works,* 95 (11): 95–6, November 1964.

70. "Maine conservationists aid solid waste disposal." *Public Works,* 98 (6): 141–42, June 1967.

71. "Making friends for landfill," part VII. *Solid Waste Management/RRJ,* 14 (1): 23, January 1971.

72. ——, part VI. *Solid Waste Management/RRJ,* 13 (10): 22, October 1970.

73. ——, part V. *Solid Waste Management/RRJ,* 13 (1): 18, January 1970.

74. "Making landfill livable." *Chemical Engineering,* p. 64, August 10, 1970.

75. Malina, J.F. and M.L. Smith. "How much refuse in your city?" *The American City,* 86 (3): 64, March 1971.

76. McElwee, W. "From landfill to streets." *The American City,* 81 (4): 24, April 1966.

77. McHenry, J.A. "How to 'doctor' a quarry for landfill." *The American City,* 85 (12): 38, December 1970.

78. McKinney, R.E. "The challenge of solid wastes research." *American Society of Civil Engineers, Journal of the Sanitary Engineering Division,* 92 (SA5): 1–6, October 1966.

79. McKinnon, J.J. "Landfill replaces controversial dump." *Public Works,* 99 (10): 121, October 1968.

80. McNulty, W. "Lower water table to solve a landfill puzzle." *Solid Waste Management/RRJ,* 12 (8): 8, August 1969.

81. Meresman, S.J. *PERT; concepts and applications to solid waste management.* Cincinnati: U.S. Dept. of Health, Education, and Welfare, 1970.

82. Merz, R.C. and R. Stone. "Progress report on study of percolation through a landfill." *Public Works,* 98 (12), December 1967.

83. ——. "Quantitative study of gas produced by decomposing refuse." *Public Works,* 99 (11): 86–87, November 1968.

84. "Methane gas explosions delay building on landfill." *Solid Waste Management/RRJ.* 12 (7): 20, July 1969.

85. "Milwaukee County settles for baling and burial." *Solid Waste Management/RRJ,* 13 (1): 6, January 1970.

86. "Model landfill contract." *Solid Waste Management/RRJ,* 13 (3): 22, March 1970.

87. Morse, N. and E.W. Roth. *Systems analysis of regional solid waste handling.* Public Health Service Publication no. 2065. Washington, D.C.: U.S. Government Printing Office, 1970.

88. *Municipal refuse disposal.* 2d ed. Chicago: American Public Works Association, 1966.

89. "National Survey of disposal needs, practices." *Solid Waste Management/RRJ,* 11 (3): 22, 46, March 1968.

90. Pagan, A.R. "Low cost pit incinerator extends life of sanitary landfill." *Public Works,* 98 (8): 131–32, August 1967.

91. "Population versus landfill." *Solid Waste Management/RRJ,* 11 (2): 46, February 1968.

92. "Railroad to carry Philadelphia waste." *Refuse Removal Journal,* 10 (10): 14, October 1967.

93. "Recommend standards for sanitary landfill design." *Solid Waste Management/RRJ,* 13 (9): 32, September 1970.

94. *Refuse collection and disposal practices.* League of Arizona cities and towns, October 1966.

95. *Refuse collection practices.* 3rd ed. Chicago: American Public Works Association, 1966.

96. Remson, I., A.A. Fungaroli, and A.W. Lawrence, "Water movement in an unsaturated sanitary landfill." *American Society of Civil Engineers, Journal of Sanitary Engineering Division,* 94 (SA2): 307–17, April 1968.

97. "Research seeks new ways to seal landfill against leaching." *Solid Waste Management/RRJ,* 14 (3): 18, March 1971.

98. Reynolds, W.F. "Abandoned strip mines studied for solid waste disposal." *Public Works,* 98 (5): 74–5, May 1967.

99. Reynolds, W. "Fairfax County's success with landfill." *American County Government,* 32 (10): 50, October 1967.

100. Riley, J. "Ski hill from garbage dump." *Parks and Recreation,* 3 (11): 37, November 1968.

101. Rose, B. "Sanitary district puts sludge to work in land reclamation." *Water and Sewage Works,* 115 (9): 393–399, September 1968.

102. Rosenberg, S. "Industrial Services of America acquires sanitary landfill acreage." *Secondary Raw Materials,* 7 (4): 62, April 1969.

103. "Sanitary fill supermechanized." *The American City,* 80 (12): 20, December 1965.

104. "Sanitary landfill: big decision for small towns." *Missouri Municipal Review,* 32 (4): 111–117, April 1967.

105. *Sanitary landfill operation agreement and recommended standards for sanitary landfill design and construction.* National Solid Wastes Management Association and U.S. Dept. of Health, Education, and Welfare, Bureau of Solid Waste Management, Report no. SW–20ts, 1969.

106. Sowers, G.F. "Foundation problems in sanitary landfills." *American Society of Civil Engineers, Journal of Sanitary Engineering Division,* 94 (SA1): 103–116, February 1968.

107. "Self-supporting landfill helps attract industry." *Public Works,* 98 (3), March 1967.

108. Smith, C.D. "A sanitary fill inside the city." *The American City,* 83 (4): 90, April 1968.

109. *Solid waste landfill stabilization; an interim report.* Ralph Stone and Company, Inc., Engineers. Cincinnati: U.S. Dept. of Health, Education, and Welfare, 1968.

110. *Solid Waste Management: A list of available literature.* U.S. Dept. of Health, Education, and Welfare, Bureau of Solid Waste Management, Report no. SW–58.8, September 1970.

111. *Solid Waste Management: A list of available literature.* U.S. Dept. of Health, Education, and Welfare, Bureau of Solid Waste Management, June 1970.

112. Sorg, T. and H.L. Hickman. *Sanitary landfill facts.* U.S. Dept. of Health, Education, and Welfare, Bureau of Solid Waste Management, Report no. SW–4ts, 1970.

113. Steiner, R.L. and R. Kantz. *Sanitary landfill; a bibliography.* Public Health Service Publication no. 1819. Washington, D.C.: U.S. Government Printing Office, 1968.

114. Stone, R. and E.T. Conrad. "Landfill compaction equipment efficiency." *Public Works,* 100 (5): 111, May 1969.

115. Stone, R., E.T. Conrad, and C. Melville. "Land conservation by aerobic landfill stabilization." *Public Works,* 99 (12): 95, December 1968.

116. Stone, R. and M. Israel. "Determining effects of recompaction on a landfill." *Public Works,* 99 (1): 72–3, January 1968.

117. "Study advocates St. Louis use nearby strip mines." *Solid waste management/Refuse Removal Journal,* 12 (4): 47, April 1969.

118. "Suburb interrupts San Francisco disposal." *Refuse Removal Journal,* 10 (1): 6, January 1967.

119. *Waste management and control.* A report to the Federal Council for Science and Technology; Committee on Pollution, Publication no. 1400. Washington, D.C.: National Academy of Sciences–National Research Council, 1966.

120. Weaver, L., ed. *Proceedings, the Surgeon General's Conference on Solid Waste Management for Metropolitan Washington.* U.S. Department of Health, Education, and Welfare, Public Health Service Publication no. 1729, July 1967.

121. "When it comes to choosing new landfill equipment." *Solid Waste Management/RRJ,* 14 (2): 18, February 1971.

122. Williams, E. "Big loader offsets burning ban." *The American City,* 85 (12): 60, December 1970.

123. Wolfe, H.B. and R.E. Zinn. "Systems analysis of solid waste disposal problems." *Public Works,* 98 (9): 99–102, September 1967.

124. Zandi, I. *Short-term sanitary landfill with guaranteed reclamation.* Philadelphia: Towne School of Civil and Mechanical Engineering, University of Pennsylvania, 1971.

125. Zausner, E. "An accounting system for sanitary landfill operations." U.S. Dept. of Health, Education and Welfare, Bureau of Solid Waste Management, Report no. SW–15ts, 1969.

Additional Bibliography

1. Anderson, J.R., and J.N. Dornbush. "Intercepting trench at landfill renders leachate into drinkable source of water." *Solid Waste Management/RRJ.* January 1972, p. 60.
2. APWA Congress. "Seminar on high pressure baling of solid waste—proceedings." Pittsburgh, Pennsylvania, September 15, 1971. Moderator: W.B. Faulkner.
3. Boyle, W.C. and R.K. Ham. "The treatability of leachate from sanitary landfills." *Proceedings,* 27th Purdue Industrial Waste Conference, Lafayette, Indiana, May 1972.
4. Coe, Jack J. "Effect of solid waste disposal on ground water quality." *Journal* of the American Water Works Association, 62 (12): 776, December 1970.
5. Emrich, G.H. "Guidelines for sanitary landfills—ground water and percolation." Presented at Engineering Foundation Research Conference on Sanitary Landfills, Deerfield, Massachusetts, August 1970.
6. Emrich, G.H. and R.A. Landon. "Generation of leachate from landfills and its subsurface movement." Presented at Annual Northeastern Regional Anti-Pollution Conference, University of Rhode Island, July 1969.
7. Ham, R.K., and J.J. Reinhardt. "Refuse milling in Europe." *Bulletin,* National Center for Resource Recovery, Inc. 3 (1): 2–11, Winter 1973.
8. Ham, R.K., W. Porter, and J.J. Reinhardt. "Refuse milling for landfill disposal." *Public Works,* published in three parts: 102 (12): 42–47, December 1971; 103 (1): 70–72, January 1972; 103 (2): 49–54, February 1972.
9. Johnson, W.C. "Gravel plus refuse = recreation." *Parks and Recreation.* September 1969.
10. Langmuir, Donald. "Controls on the amounts of pollutants in subsurface waters." *Earth and Mineral Sciences,* 42 (2): 9, November 1972.
11. Salvato, J.A., W.G. Wilkie, and B.E. Mead. "Sanitary landfill-leaching prevention and control." *Journal,* Water Pollution Control Federation, 43 (10); 2084, October 1971.
12. Steiner, R.L., A.A. Fungaroli, R.J. Schoenberger, and P.W. Purdom. "Criteria for sanitary landfill development." *Public Works,* 102 (3): 77, March 1971.
13. Zanoni, A.E. "Ground water pollution from sanitary landfills and refuse dump grounds—a critical review." Research Report 69, Department of Natural Resources, Madison, Wisconsin, 1971.

115

Index

117